Integrated Catastrophe Risk Modeling

CU01510079

Advances in Natural and Technological Hazards Research

Volume 32

For further volumes:
http://www.springer.com/series/6362

Aniello Amendola • Tatiana Ermolieva
Joanne Linnerooth-Bayer • Reinhard Mechler
Editors

Integrated Catastrophe Risk Modeling

Supporting Policy Processes

 Springer

Editors

Aniello Amendola
Risk, Policy and Vulnerability
  (RPV) Program
International Institute for
  Applied Systems Analysis (IIASA)
Laxenburg, Austria

Joanne Linnerooth-Bayer
Risk, Policy and Vulnerability
  (RPV) Program
International Institute for
  Applied Systems Analysis (IIASA)
Laxenburg, Austria

Tatiana Ermolieva
Ecosystems, Services and Management
  (ESM) Program
International Institute for
  Applied Systems Analysis (IIASA)
Laxenburg, Austria

Reinhard Mechler
Risk, Policy and Vulnerability
  (RPV) Program
International Institute for
  Applied Systems Analysis (IIASA)
Laxenburg, Austria

ISSN 1878-9897   ISSN 2213-6959 (electronic)
ISBN 978-94-017-8498-6   ISBN 978-94-007-2226-2 (eBook)
DOI 10.1007/978-94-007-2226-2
Springer Dordrecht Heidelberg New York London

Preface

This book aims to advance risk management policy and its implementation by demonstrating the application of novel techniques, including integrated catastrophe models, to aid policy decisions on contemporary disaster risk issues.

With the dramatic rise in disaster events across an increasingly populous and interdependent world, people and communities are recognizing the importance of reducing their human and economic toll. Efforts are thus being intensified to manage risk and incorporate risk management principles into policymaking.

Science-based risk management policy is not without its challenges. Given the dynamic demographic, economic, and social context in which most hazards are embedded, including the changing climate, it is not only difficult to assess risk, but standard statistical measures are inappropriate for high-impact, low-probability events with probability distributions characterized by fat tails. In addition, data on rare events are, by definition, very limited and fraught with uncertainties. Beyond the assessment of risk itself, identifying robust policy options in a highly uncertain future world poses equally difficult challenges. Finally, experts alone cannot evaluate policy options, as these depend on the values and preferences of those affected. This raises the challenge of designing and assisting stakeholder processes that can inform risk management decisions.

In this book, we address all these challenges by developing and applying modeling techniques for the assessment and management of catastrophe risk through an integrated "systems" approach. We emphasize integration across natural and social systems, applying models that take account of the intensity and frequency of natural phenomena combined with the exposure and vulnerability of social and economic systems. Integration also means gauging the complex interdependencies of risk across different temporal and spatial scales, which often requires estimates to be made into the distant future and at the local, regional, and even global levels. Integration, too, means taking due consideration of the manifold uncertainties and social constructions of disaster risk. Finally, and perhaps most importantly for the risk management policy process, integration means listening and responding to the plural and competing values and worldviews of stakeholders and policymakers.

Catastrophe models are an important part of integrated assessment, as they explore the drivers of disasters (hazard, exposure, and vulnerability), simulate future events based on historical data and expert judgment, apply appropriate statistical distributions, and take account of future drivers, including climate change. As well as providing risk estimates, as this volume shows, models can be embedded in support systems that can account for conflicting values and for the views of multiple stakeholders. In so doing, they provide useful knowledge and support to risk management policies.

The uniqueness of this book lies in its usefulness to real-world policies on catastrophe risk management and in the novelty of the approaches used for this purpose. The three parts of this volume begin with general discussions of catastrophe models for informing risk management policies, and then turn to the implications of disaster risk for economic growth and socioeconomic development along with associated options for managing risk. Finally, we focus on the Tisza River basin in Hungary, describing the implementation of a model-based stakeholder process for managing flood risk in this area.

Specific applications, among others, include designing insurance strategies for seismic risk in Italy; assessing strategies for managing flood risk in Austria and northern Vietnam; evaluating large infrastructure projects throughout the world; examining the development implications of extreme climate events in Nepal; developing catastrophe bonds for public sector risk management in Mexico; informing the development of the Caribbean Catastrophe Risk Insurance Facility; and implementing a model-based participatory process for managing flood risks in Hungary. The applications break new ground by applying advanced modeling techniques to the policy issues at hand. Methodological innovations include novel stochastic optimization approaches and probabilistic risk estimation taking account of indirect losses and climate change. The applications are also innovative in that they are designed for user-friendly policy support. This, as the research shows, can prove to be instrumental in helping stakeholders holding strongly divergent views reach policy consensus and in helping national policymakers, donors, and development bankers devise risk financing strategies for implementation in highly vulnerable developing countries.

The research was carried out by scientists and their collaborators at the International Institute for Applied Systems Analysis (IIASA), which conducts policy-oriented systems research into problems that are too large or too complex to be solved by a single academic discipline. The research is also linked with other international institutions such as the Intergovernmental Panel on Climate Change, UN development agencies, the European Commission, and international finance institutions such as the Asian Development Bank, Inter-American Development Bank and the World Bank. The researchers represent a wide range of physical and social science disciplines, including mathematics, statistics, systems modeling, geology, meteorology, hydrology, physics, engineering, computer sciences, economics, decision analysis, and sociology. The case studies would have not been feasible without the support and availability of data from the national or local institutions involved, as acknowledged in the relevant chapters. In particular, we are grateful to the Swedish Research Council for Environment, Agricultural Sciences and Spatial Planning (FORMAS) for providing funding to IIASA for the Hungarian Tisza river project.

It is our hope that this volume will contribute positively to the design and implementation of scientifically grounded and socially acceptable policy options able to reduce the unacceptable human and economic toll of natural hazards, today and in the future.

Aniello Amendola
Tatiana Ermolieva
Joanne Linnerooth-Bayer
Reinhard Mechler

stakeholders with conflicting objectives, who are participating in a process charged with negotiating among different policy options. An example is the design of a national insurance system. In both policy contexts, models can be an integral part of a support or optimization tool.

1.5   Book Overview

The chapters in this book are grouped into three parts:

- Catastrophe Models for Informing Risk Management Policy
- Disasters and Growth: Modeling and Managing Country-Wide Catastrophe Risk
- Tisza River Basin in Hungary: Flood Risk Management, Multi-stakeholder Processes and Conflict Resolution

1.5.1   Part I: Catastrophe Models for Informing Risk Management Policy

Since it is methodologically intractable to optimize decisions in the face of the large uncertainties inherent in low-probability events, the research reported in Part I shows how the analyst can specify solutions that are "robust", that is, that are best given multiple uncertain outcomes. Applications include the design of insurance strategies for seismic risk, risk management strategies for flash and river flooding, and incorporating the discount rate in decisions on river dams.

A major methodological challenge addressed in Part I is how to optimize a policy decision in the face of a very large number of catastrophe scenarios. Since a strategy that is optimal given one scenario may not be optimal overall, the question is how to design interventions that are robust with respect to all potential outcomes. For this purpose, the authors develop and apply an innovative methodology that combines catastrophe simulation models with stochastic optimization techniques to enable the simultaneous analysis of complex interdependencies among damages at different locations and robust structural and financial risk reducing measures.

More details on the chapters in Part I are discussed in Sect. 1.6.

1.5.2   Part II: Disasters and Growth: Modeling and Managing Country-Wide Catastrophe Risk

The impact of disasters on aggregate economic performance and development, and measures to increase country-wide resilience to catastrophes, are the subjects of Part II of this volume (for an introduction see Chap. 6). The chapters reflect on how catastrophic shocks impact on economic performance. Even in the case of well-behaved economic models, as Chap. 7 shows, shocks may modify the structure of the economy and lead to traps and thresholds triggering stagnation and shrinking. IIASA's catastrophic simulation model (CATSIM), which is the basis of the chapters in this section, simulates public sector losses resulting from external shocks. CATSIM uniquely incorporates an economic growth model to include the indirect impacts. The model has proved useful in planning financial management interventions.

CATSIM applications discussed in Part II are targeted to highly vulnerable developing countries, and include:

- the development of catastrophe bonds for public sector risk management in Mexico, for which CATSIM provided a picture of the different layers of risks posed by earthquakes to the public finances and helped identify which risks could be transferred to the international market at an acceptable cost (Cardenas et al. 2007);
- the design of the Caribbean Catastrophe Risk Insurance Facility (CCRIF), where CATSIM provided support for a stakeholder workshop in Barbados involving nearly all Caribbean country finance ministries (Hochrainer and Mechler 2009); and
- the design of a global climate-related disaster fund, for which CATSIM provided the quantitative basis for its implementation.

1.5.3 Part III: Tisza River Basin in Hungary: Flood Risk Management, Multi-stakeholder Processes and Conflict Resolution

The third part of this book shows how a system analytical approach can provide insights in a multi-stakeholder participatory process. Specifically, the authors report on a 3-year stakeholder process (designed and implemented by IIASA researchers) that ultimately produced a consensus on a nation-wide flood insurance system for Hungary. A flood catastrophe model for the Upper Tisza was instrumental for the success of this process. The hydrology-based model demonstrated the costs of flooding on the Upper Tisza, and how these costs would be distributed differently among residents, insurers and the government depending on the design of the national insurance system. In accompanying chapters, the authors address the elements of vulnerability in the Upper Tisza region, and also develop a methodology for including climate change in the model.

Parts II and III of this book are introduced in Chaps. 6 and 10, respectively; in what follows we briefly introduce Part I with an introduction and short description of Chaps. 2, 3, 4, and 5.

1.6 Part I: Catastrophe Models for Informing Risk Management Policy

1.6.1 Introduction

Part I discusses the theory and application of integrated catastrophe models, including:

- the management of flash flood risk in Vienna, Austria;
- modeling catastrophe risk in the design of an insurance system for Hungary;
- handling multiple criteria in flood risk management in Vietnam; and
- incorporating the discount rate in decisions on river dams.

This research is based on earlier work at IIASA, including models related to the design of earthquake insurance policies in Russia and Italy by integrating an earthquake hazard module and GIS-based maps of seismic intensities (Petrini 1995) and vulnerabilities (Amendola et al. 2000a, b).

A challenge explicitly addressed in Part I is evaluating the interdependencies of hazard scenarios, risk management strategies and their outcomes. In conventional catastrophe models, the assessment of policies are typically performed in a scenario-by-scenario (what-if) manner; yet, the evaluation of all interdependencies is methodologically intractable if there are large numbers of alternative scenarios and policy combinations. A strategy optimal against one scenario may not be optimal against multiple scenarios. This has prompted authors in this volume to develop methodologies that make possible the identification of strategies that are *robust* with respect to all plausible catastrophe scenarios. By *robust* we mean that the strategies satisfy pre-specified conditions important to decision outcomes, which requires combining simulation methods and stochastic optimization. Robust solutions are based on the reality that exact evaluations of optimal decisions can be intractable; yet by combining simulation methods and stochastic optimization it is feasible to identify solutions that satisfy pre-specified constraints (Ermoliev and Wets 1988). In Part I of this book, we not only demonstrate this innovative "combination" methodology, but also show how the constraints can be sensitive to the preferences of multiple stakeholders.

When multiple dimensions of the decision space are considered, the search for robust solutions becomes more complex. Outcomes of the policy decision might include such incommensurable elements as protecting livelihoods, biodiversity and cultural heritage, which cannot easily be combined into a single monetary valuation. Multi-criteria decision support may be called for, and this section demonstrates how to combine catastrophe models and multi-criteria decision support.

1.6.2 Chapter Descriptions

The first chapter in Part I by Compton et al., *Modeling Risk and Uncertainty. Managing Flash Flood Risk in Vienna,* presents an application of catastrophe modeling to flash flood risk management in Vienna, where the policy options include both risk reduction and insurance. The study makes use of hydrological flood risk assessment and simulation modeling. The analysis demonstrates that the largest damages to be expected are for the city subway system.

The methodology is particularly innovative in that it analyzes selected policies in the presence of major uncertainties, both aleatory and epistemic. Facing these uncertainties especially under limited financial resources, the authors emphasize the need for a combination of structural risk prevention and financial risk sharing measures. The study is also innovative in its combination of Monte Carlo simulation with stochastic optimization (see also Chaps. 3 and 15 of this book). The risks are estimated and presented in the form of a "risk curve" or CCDF (complementary cumulative distribution function), which plots the magnitude of an event on the horizontal axis vs. the probability of exceeding that magnitude on the vertical axis (similar to the probability of exceedance discussed earlier). The epistemic uncertainty is represented by error bands of any desired confidence level that surround that curve.

A major finding of this chapter is that although structural (loss-preventing) and financial (loss-spreading) policy measures may have significantly different characteristics, they can be examined in a consistent and unified approach if an appropriate measure of risk is identified. By combining ex ante financial provisions, such as a reserve fund and insurance, with prevention measures, such as flood retention basins, an overall risk management strategy can be identified that decreases total costs and reduces the likelihood of catastrophic financial losses and their uncertainties.

Chapter 2 by T. Ermolieva and Y. Ermoliev, *Modeling Catastrophe Risk for Designing Insurance Systems,* presents an integrated catastrophe management model for the analysis of structural and financial measures reducing the impacts of catastrophes on regional welfare. The authors discuss the challenging institutional problems related to designing regional programs for insuring rare high-impact risks. Their model addresses the characteristics of catastrophic risks: highly mutually dependent and spatially distributed endogenous risks, uncertainty in available data and models, the need for robust strategies in long term perspectives, and explicit modeling of constraints and expectations of involved agents and stakeholders (such as households, firms, local and central governments, insurers and investors). The model consists of four modules: hazard, vulnerability, a multi-agent accounting system, and a stochastic optimization module for determining optimal prevention and financing policies. The chapter describes the theoretical concepts and applies them to a case study focusing on insurability of seismic risk in the Tuscany region, Italy.

For this purpose, the integrated risk management model explicitly incorporates a hazard simulation module, i.e., an earthquake generator that accounts for the geological characteristics of the region as well as the geographical distribution of exposed buildings and their vulnerability. The optimization is based on reasonable assumptions about the aversion of insurers to the risk of insolvency and the desire of the insured not to overpay insurance premiums. It shows that it is possible to design optimal insurance strategies that are robust with respect to loss dependencies and uncertainties.

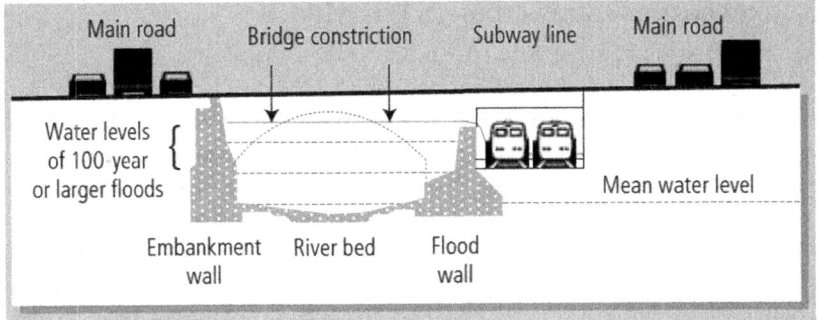

Fig. 2.2 The threat of Vienna River to the subway system

constructed in an open section at the right river bank, and the main roads on both sides (see Fig. 2.2). For 7.5 km, it is situated mostly in open sections beside the river before it enters the underground track. A partition wall protects the subway line from floods. Portable flood barriers can be installed in two locations in order to prevent the overflowing water amounts from being conveyed to underground sections of the line that include major subway junctions. These emergency measures were installed recently and require a 6-h lead time for installation. At the left embankment main roads are located, together with densely populated areas. Various service pipes are placed under the road embankments.

Several failure mechanisms that could lead to severe damage to the subway have been analyzed. The term overtopping is used for a situation where the mean water level is higher than the wall crest. This contrasts with "wave overtopping", which refers to the temporal and spatial oscillations of the water surface over the flood wall. Although no past inundation or other flood damage to the subway or the embankment has been reported, it is generally agreed that flood risk has increased during the last decades due to development of residential areas and other land use changes in the catchment. In case of intensive overflowing and the absence or malfunction of the transverse portable flood barriers, the U4 subway line acts as flood bypass conveying water downstream to the tunnels and nearly all connected lines are inundated.

Another failure mechanism is seen in the collapse of the subway partition wall and/or embankment walls, either initiated by local scouring or a high water pressure exceeding the resistance of the separation wall.

At the time of the study (March 2003), the situation of the flood protection system in the Vienna River basin was characterized by a sequence of partly upgraded detention reservoirs, local protection measures, such as water proofing of some objects along the river bed and fortification of the separation wall. The hydrologic investigations in this study distinguish among several construction and operational states of the retention basin system:

1. Hypothetical natural state without any artificial retention capacity;
2. Reservoir state before beginning of the upgrading works in 1997;
3. Recent (2002) state;
4. Reservoir state after completed upgrading.

2.5 Stochastic Optimization (STO) Model

Catastrophe models have been developed at IIASA since 1997 with reference to catastrophe insurance issues (Ermolieva et al. 1997, Emoliev et al. 2000; Amendola et al. 2000a, b, 2005, see also Chaps. 3, 7, 15 by Ermolieva et al. in this book). Catastrophe models offer a natural setting for analyzing a series of loss preventive measures in addition to loss spreading measures under consideration of uncertainty. A common element in most catastrophe models is the use of decomposition, a staple element in systems-analytic thinking (Bier et al. 1999). Decomposition is implemented by the creation of modules or submodels Indeed, they include a "hazard" sub-model driving the risk, a "loss" estimation sub-model and a "management" sub-model examining the impact of different decisions (Walker 1997; Clark 2002). They produce outputs that are distributional, since they make use of simulations in each of which the parameters of the submodels can be changed by Monte Carlo techniques within their uncertainty distributions.

For this study the hazard module was developed after a hydrologic and hydraulic analysis to come up with an estimate of the frequency of failure of the investigated protection system and to give an approximation of the severity of a failure event. Uncertainties in the input data are processed by Monte-Carlo methods.

For modeling the watershed hydrology, the rainfall depth that was sampled for a defined return period is transferred into a peak discharge by stepwise deterministic relations for different constructional and operational detention reservoir states. These transfer functions were derived from rainfall-runoff models for the rural and the urban river reach. Several basic random variables were introduced in the Monte-Carlo simulations in order to incorporate uncertainties in the channel roughness, the river cross-section station and elevation and the energy loss due to bridge constrictions. From the output of each hydraulic model-run, the occurrence of several possible flood-induced failure modes was evaluated. These failure types comprise overflowing, structural damages like tipping of a flood wall, scouring of the river bed and collapsing of river bank structures.

For the failure assessment, a Monte Carlo approach is applied by varying randomly basic hydraulic parameters, such as data about cross section geometry, roughness coefficients and the erodibility of the river bed, within a predefined range. As a consequence, all the basic hydraulic variables can be characterzied by their expectation values as well as by their range. The water pressure on the flood wall, the critical river bed material's shear stress and the scour depth, the partly blocked flow profile due to collapsed bank structures and backfill material can be considered as random variables. The storm depth is modeled by a Gumbel distribution. Finally, storm depths are transferred into discharge rates via rainfall-runoff models. Models, data and uncertainties are discussed extensively in the quoted full report by Compton et al. (2009).

The hydrologic/hydraulic simulations for the conditional probability of failure are indicated in Fig. 2.3, which cover scenarios of 12 return periods for the current and the projected state of the flood control reservoirs. The total probability

Fig. 2.3 Probability of failure (PF) for the current (*above*) and the projected state of the flood control reservoirs, conditional to a flood of return period Tr. *Crosses* denote simulated data points; *curves* are fitted to obtain continuous distributions to be used in the simulation

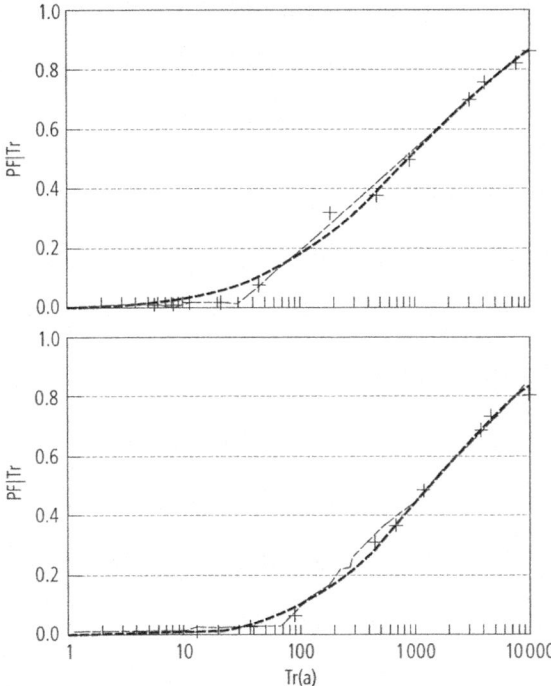

concept (e.g., in Ang and Tang 1975; Plate 1993) is used for the integration of all conditional probabilities weighted by their occurrence probability and gives an estimate of the probability that the system fails in 1 year. By using the fitted lognormal distribution in Fig. 2.3, the probability of failure in a year is calculated to diminish from a value of 4.9 10^{-3} without a retention basin to 2,5 10^{-3} for the projected basin state. The data however are very sensitive to the curve fitting procedures and such uncertainty needs to be taken into account in the simulations.

The basic approach to estimating the flood damage during flooding of the subway is to presume that there is a proportionality relationship between the length of track flooded and the resulting direct damages (Neukirchen 1994), with modifying factors taking into account the magnitude of the rate of overflowing water (Q). Parameters have been estimated after analysis of subway flooding incidents that have occurred in other countries, for which data have been summarized in Table 2.1 and by decomposing the subway system into major systems (e.g., track, communication systems, power systems, etc.) and evaluating the percent damage to the different systems as a result of inundation.

The parameters α and λ have been estimated for DD, that express proportionality to the damaged length (L) and dependence on overflowing (Q) respectively:

$$DD = \alpha\left(1 - e^{-\lambda Q_{overflowing}}\right)L.$$

Table 2.1 Reported damages in subway flooding incidents

	Boston 1996	Seoul 1998	Taipei 2001	Prague 2002
Total system cost (€m)	Not available	790[a]	15,000[b]	Not available
Total construction cost per km (€m)	Not available	18	~180	Not available
Km/track flooded	2–3	11	9–12	15–20
Amount of water (thousand m^3)	53	800	Not available	>1,000
Reported flood damage	~10	40	60–140	66–240
Computed damage per km	1.3–4	~3.6	0.9–12	4.44–16

[a]Line 7 only
[b]Entire system (86 km)

The connected uncertainty is quite large as based on a limited data set, which in addition has been derived from manifold and not always equally reliable sources, such as insurance companies, national agencies and press releases. The interested reader can find in Compton et al. (2009) a detailed analysis and a rather extensive reference set.

2.6   Risk Management Measures and Integrated Modeling

The basic approach to developing the risk curve can be subdivided into five steps:

Step 1: The planning period of interest (PPI) corresponding to the time frame of concern of the decision maker is identified;

Step 2: A set of rainfall-probability pairs is generated by assuming that chosen severe storms of arbitrary magnitude occur within the PPI and evaluating their likelihood as based upon the assumed Gumbel distribution;

Step 3: By using the chosen rainfall-runoff relationship, the storm rainfall is transformed in the water discharge into the Vienna river. It is then possible to determine the conditional likelihood and magnitude of system failure for different levels of runoff and the corresponding amounts of water entering the subway system. In this way a set of rainfall-probability pairs is transformed to a set of overflowing water-probability pairs;

Step 4: The direct tangible damages resulting from overflowing water is determined, so that the set of rainfall-probability pairs is transformed to a set of direct damage-probability pairs. By plotting the sets of damage/probability pairs on the risk curve described previously, a scatterplot is generated if parameter values are sampled from distributions representing epistemic uncertainty in the preceding calculations. To provide a clear representation of the relationships, rather than scatterplots, curves are produced by taking subsets corresponding to specified probability intervals and computing the mean or fractiles of the distributions. This results in a conditional probability distribution representing the epistemic uncertainty in damages given that an event falling within a specified probability band (e.g., the 100 year flood) occurs;

Step 5: It is now possible to estimate the impact of non-structural mitigation measures, such as insurance or reserve funds, on the total costs (pre- and post disaster) incurred to manage the flood risk. This is done by estimating the extent to which the losses can be compensated from a reserve fund or an insurance policy, and if the losses cannot be fully covered, asking what loan is needed to cover the costs. The premiums paid before the event are counted as costs, as are the interest payments made on any loans taken out after the event.

These steps are described in detail by Compton et al. (2009). In the following a brief discussion is devoted to the financial module.

Direct damages are the input to the financial module.[1] Two ex-ante financing measures, insurance and a reserve fund, are considered. One ex-post financing measure, borrowing, is also considered.

An important parameter in examining the impact of different financial measures, such as insurance or reserve funds, is the arrival time of the first event, as this will determine to what extent a reserve fund has accumulated funds or for how long premiums have been paid. The arrival time can be modeled as a uniform distribution.

Insurance can be simulated as either proportional insurance or as excess of loss insurance, or both. However, in the model it is currently possible to define only one layer. The following parameters are used to characterize insurance: the attachment point, or "deductible", of the insurance (100% of all losses below the attachment point are borne by the policyholder); and the proportion of losses within the insured layer that is borne by the policyholder. Setting the latter to 1 causes insurance to be inactive (i.e., if the policyholder bears 100% of the losses, then the insurer pays no claims).

The claims are computed as a proportion of the total loss exceeding the attachment point. However, in order to define the upper limit of the layer, the claim payments are capped by an exit point, which is the maximum claim payment by the insurer.

The accumulated insurance reserve is simply the accumulated premiums minus the claim payment at the time of the catastrophe. Determination of the premium takes into account a loading factor for low probability events. If collected premiums are sufficient to cover the claims, the insurance reserve is positive and the premiums have been "overpaid". If the collected premiums are insufficient to cover the claims, then the insurance reserve is negative and the claims are "underpaid". The losses retained by the policyholder are simply the damages minus the claims.

We presume that the reserve fund is invested in a relatively safe security, such as bonds. The reserve fund comprises two components: a one-time initial investment and a constant annual payment.

The difference between the contribution and the balance represents the benefit of the reserve fund. It can be seen that the benefit is quite small for short time horizons

---

[1] Computation of financial parameters follows Mechler et al. (2006), Hochrainer (2006), see also Chap. 8 in this book. For a discussion of costs of reserve funds see Kielholz (2000).

(<10 years, but increases significantly thereafter due to compounding). Because we have chosen to integrate financial uncertainties with structural uncertainties, we have modeled the yield of the reserve fund as a random variable.

Post-disaster borrowing is considered making use of an extremely simple model. The cost of a loan is simply the difference between the amount borrowed and the amount repaid. We take the period to be fixed at 30 years and assume an average loan interest rate of 4% (real). We assume that the interest rate is an uncertain random variable that can range between 2 and 6% real at a 2σ (95.4%) confidence level. These values are chosen somewhat arbitrarily but are intended to emphasize the fact that borrowing is also a mitigation measure with substantial costs, and that the decision not to mitigate may be an implicit decision to assume a loan at whatever terms may be obtainable if a disaster occurs.

2.7   Summary of the Results

The richness of the achievable results is demonstrated by the cost distribution in Figs. 2.4, 2.5, and 2.6. Figure 2.4 shows (in the ordinate) the probability of exceeding the total cost given in the abscissa for the different structural measures considered. The dotted lines are the upper bound (90% confidence level) and the lower bound (10% confidence level) given the assumed epistemic uncertainties. A "no-action" alternative was considered to establish a base case. It is assumed that if damages occur, the losses will be covered by a loan. Alternative 1 is the installation of a portable flood barrier at the openings to the covered sections of the metro. Alternative 2 comprises upgrading of the basins to allow controlled filling and release of floodwaters. This system, coupled with a real-time flood forecasting system, is currently being installed to increase the level of protection

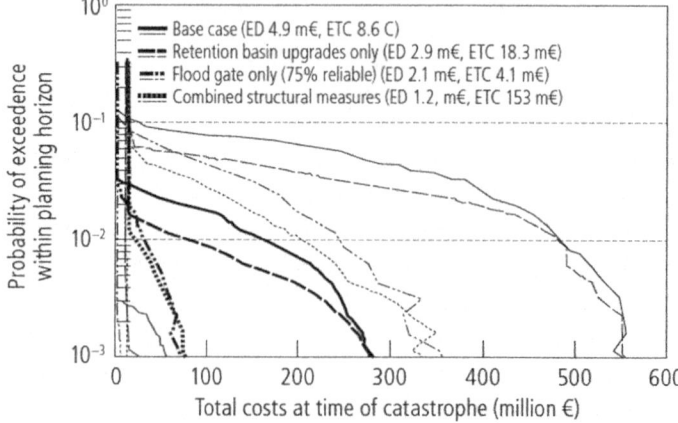

Fig. 2.4 Total costs for different structural alternatives

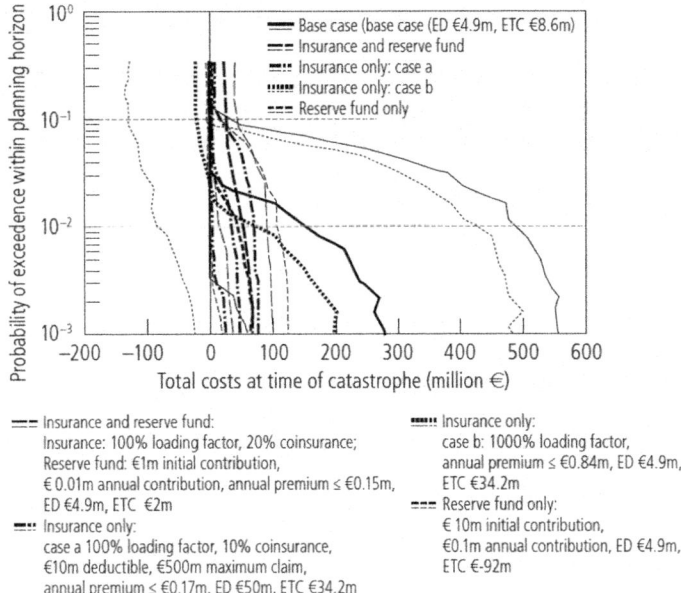

== Insurance and reserve fund:
Insurance: 100% loading factor, 20% coinsurance;
Reserve fund: €1m initial contribution,
€ 0.01m annual contribution, annual premium ≤ €0.15m,
ED €4.9m, ETC €2m
=:: Insurance only:
case a 100% loading factor, 10% coinsurance,
€10m deductible, €500m maximum claim,
annual premium ≤ €0.17m, ED €50m, ETC €34.2m

⬛⬛:: Insurance only:
case b: 1000% loading factor,
annual premium ≤ €0.84m, ED €4.9m,
ETC €34.2m
=== Reserve fund only:
€ 10m initial contribution,
€0.1m annual contribution, ED €4.9m,
ETC €-92m

Fig. 2.5 Total costs for different financial measures

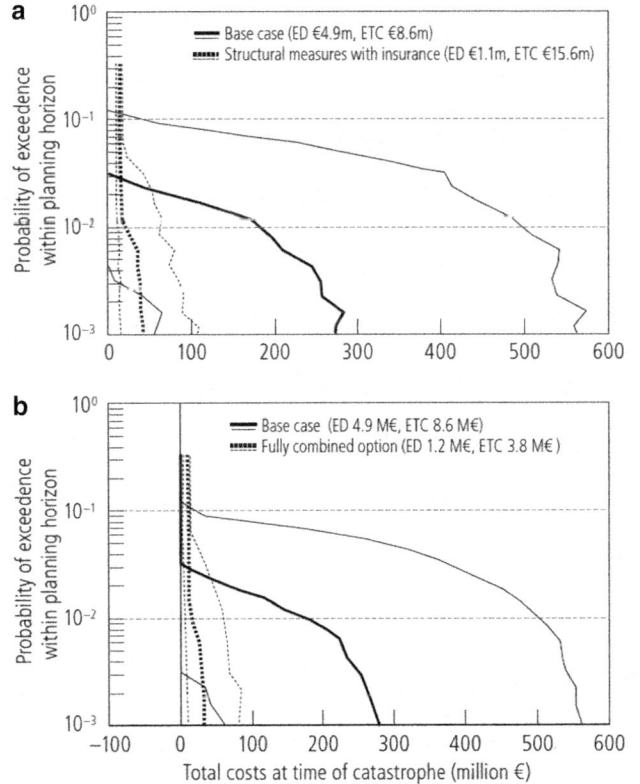

Fig. 2.6 Total costs for fully integrated financial and structural measures

against extremely rare floods. The combined scenario represents the combination of portable flood barriers and detention basin upgrades.

Figure 2.5 shows the cost mitigation effects of different financial measures, including the existence of a reserve fund, two different insurance schemes and combination of insurance and the reserve fund.

Finally, Fig. 2.6 shows the total cost reduction obtained by fully combining structural and financial measures.

2.8   Some Conclusions

Our primary conclusion is that the implementation of a concept of risk that integrates the different technical perspectives on risk into a unified framework is feasible and yields valuable insights into the nature of the protection provided by different mitigation alternatives. This implementation of an integrated concept of risk is achieved by identifying a clear assessment variable (total ex-ante and ex-post costs of mitigating flood damage) and expressing the probability distribution of this variable under different mitigation scenarios using a stochastic complementary cumulative distribution function, or "risk curve". This approach provides considerable additional relevant information to a decision maker. It also allows the decision maker to structure the problem in such a way as to obtain a clearer indication of the advantages and disadvantages of different mitigation options. This has been demonstrated by examining a current problem faced by decision makers and using, to the maximum extent possible, accurate and relevant data. We further note that the results highlight the fact that the advantages and disadvantages of a particular proposed mitigation option are complex and cannot always be reduced to a single-valued metric, such as expected benefit or system reliability, as is typical of an actuarial approach and a probabilistic approach, respectively.

A second finding of the study is that although structural (loss-preventing) and financial (loss-spreading) mitigation measures may have significantly different characteristics, they may still be examined in a consistent way if an appropriate measure of risk can be identified. This is closely connected with the use of a broader conception of risk that identifies the strengths and weaknesses of different mitigation measures. Understanding the comparative strengths and weaknesses of different instruments can assist in the design of a system in which the advantages of some measures are used to offset the disadvantages of other measures, thereby reducing and controlling the risks.

As our example showed, the explicit treatment of epistemic and aleatory uncertainty allowed the analysis to clarify the different characteristics of reserve funds vs. insurance. In this case, the reserve fund served to reduce (or even offset) the cost of ex-post borrowing, although it provided essentially no protection against very large events and did not reduce the uncertainty in the loss curve. The effect of the reserve fund was to shift the risk curve in a beneficial direction at all probability levels. On the other hand, insurance provided protection against the relatively larger and less likely losses and reduced the uncertainty associated with the large events. The effect of the floodgate was similar to that of insurance in that losses from very rare events were reduced; however, insurance was clearly more effective at reducing the uncertainty of large losses, at the expense of increasing costs. Both of these were quite different from the type of protection provided by the detention basins, which served to reduce the probability of losses but was subject to considerable uncertainty about the losses when the capacity of the basins could be overwhelmed by beyond design-basis storms. The synergistic effects of combined measures were apparent, in that the use of structural measures assisted in mitigating the major drawback of insurance (the high cost) by reducing expected losses while the insurance policy managed the residual uncertainty associated with the structural measures. Also, the effect of a reserve fund was enhanced when combined with loss reduction techniques that extended the potential for accumulating adequate reserve funds. In this case, we were able to demonstrate that using plausible values and realistic options drawn from a real flood risk management problem, considerable reduction in the total cost of mitigating flood damage may be achieved by combining structural measures with financial measures.

References

Amendola A (2001) Recent paradigms for risk informed decision-making. Saf Sci 40(1–4):17–30
Amendola A, Ermoliev Y, Ermolieva T (2000a) Earthquake risk management: a case study for an Italian region. In: Proceedings of the second Euroconference on global change and catastrophe risk management: earthquake risks in Europe. International Institute for Applied Systems Analysis, Laxenburg, Austria, 6–9 July 2000
Amendola A, Ermoliev YM, Ermolieva TY, Gitis V, Koff G, Linnerooth-Bayer J (2000b) A systems approach to modeling catastrophic risk and insurability. Nat Hazard 21:381–393
Amendola A, Ermolieva T, Ermoliev Y (2005) Catastrophe risk management – vulnerability and equity. In: Gheorghe AV (ed) Integrated risk and vulnerability management assisted by decision support systems – relevance and impacts on governance. Springer, Dordrecht, pp 383–401
Ang A, Tang W (1975) Probability concepts in engineering planning and design. Wiley, New York
Bier VM, Yacov Y, Haimes JH, Lambert NC, Zimmerman M & R (1999) A survey of approaches for assessing and managing the risk of extremes. Risk Anal 191:83–94
BMLFUW (2002) Gewässerschutzbericht. Bundesministerium für Land- und Forstwirtschaft, Vienna, Austria, ISBN: 3-85 174-042-4
Clark KM (2002) The use of computer modeling in estimating and managing future catastrophe losses. Geneva Pap Risk Insur 27:181–195
Compton KL, Faber R, Ermolieva T, Linnerooth-Bayer J, Nachtnebel HP (2009) Uncertainty and disaster risk management. Modeling the flash flood risk to Vienna and its subway system. International Institute for Applied Systems Analysis, IIASA report RR-09-002, Laxenburg, Austria
Covello VT, Merkhofer MW (1994) Risk assessment methods: approaches for assessing health and environmental risks. Plenum Press, New York
Emoliev Y, Ermolieva T, MacDonald G, Amendola A (2000) A systems approach to catastrophe management. Eur J Oper Res 122:452–460

Ermolieva T, Ermoliev Y, Norkin V (1997) On the role of advanced modeling. In: Drottz-Sjöberg BM (ed) Managing catastrophic risks, Proceeding of the new risk frontiers: conference for the 10th anniversary of the society for risk analysis – Europe, The Center for Risk Research, Stockholm, pp 68–74

European Commission EC (2000) Communication from the Commission on the precautionary principle, COM(2000)1. Commission of the European Communities, Bruxelles

Faber R, Nachtnebel HP (2003) Project report of the Universität für Bodenkultur (BOKU) and the Institut für Wasserwirtschaft, Hydrologie und Konstruktiven Wasserbau (IWHW), Vienna, Austria

Hochrainer S (2006) Macroeconomic risk management against natural disasters. Deutscher Universitätsverlag (DUV), Wiesbaden

Kielholz W (2000) The cost of capital for insurance companies. Geneva Pap Risk Insur 25:4–24

Knight FH (1921) Risk, uncertainty, and profit. Hart, Schaffner & Marx; Houghton Mifflin Company. Download at: http://www.econlib.org/library/Knight/knRUP1.html. Accessed Jan 2012

Konecny F, Nachtnebel HP (1985) Extreme value processes and the evaluation of risk in flood analysis. Appl Math Model 9:11–16

Mechler R, Linnerooth-Bayer J, Hochrainer S, Pflug G (2006) Assessing financial vulnerability and coping capacity: the IIASA CATSIM model. In: Birkmann J (ed) Measuring vulnerability and coping capacity to hazards of natural origin: concepts and methods. United Nations University Press, Tokyo, pp 380–398

Nachtnebel HP (2000) Flood Control Measures and Strategies for Risk Mitigation. In: Gillet F, Zanolini F (eds) Risque naturels en montagne. Actes de Colloque, Editions Cemagref, Grenoble. 12–14 Apr 1999, pp 387–389.

Neukirchen H (1994) Hochwasserr Rückhalteanlagen für den Wienfluss: Kosten Nutzen Untersuchung. Project for the Municipal Hydraulic Engineering Department MA, Vienna, Austria

Plate EJ (1993) Statistik und Angewandte Wahrscheinlichkeitslehre für Bauingenieure. Ernst & Sohn, Berlin

Renn O (1992) Concepts of risk: a classification. In: Krimsky S, Golding D (eds) Social theories of risk. Praeger Publishers, Westport, pp 53–82

Walker G (1997) Current developments in catastrophe modelling. In: Britton NR, Oliver J (eds) Financial risk management for natural catastrophes. Griffith University, Brisbane, pp 17–35

Chapter 3
Modeling Catastrophe Risk for Designing Insurance Systems

Tatiana Ermolieva and Yuri Ermoliev

Abstract In catastrophe management, risk spreading is one of the important measures for increasing societal resilience to disasters. In this paper we discuss an integrated catastrophe management model which explores alternative risk spreading options. As a case study we consider the seismic prone Tuscany region of Italy. Special attention is given to the evaluation of a public loss-spreading program involving partial compensation to victims by the central government and the spreading of risks through a pool of insurers on the basis of location-specific exposures. GIS-based catastrophe models and stochastic optimization methods are used to guide policy analysis with respect to location-specific risk exposures. The use of economically sound risk indicators lead to convex stochastic optimization problems strongly connected with nonconvex insolvency constraint and Conditional Value-at-Risk (CVaR).

Keywords Flood and seismic risk • Catastrophe modeling • Catastrophic insurance • Contingent credit • Stochastic optimization • Safety constraints • Risk measures

3.1 Introduction[1]

Losses from human made and natural catastrophes are rapidly increasing (Munich 2009, 2011a, b). The main reason for this is the clustering of people and capital in hazard-prone areas as well as the creation of new hazard-prone areas (Dilley

---

[1] The chapter is a modified version of a previous paper, which is reprinted with permission of the Society for Industrial and Applied Mathematics "Copyright (c) 2005 Society for Industrial and Applied Mathematics. All rights reserved."

T. Ermolieva (✉)
Ecosystems, Services and Management (ESM) Program, International Institute for Applied Systems Analysis (IIASA), Schlossplatz 1, A-2361 Laxenburg, Austria
e-mail: ermol@iiasa.ac.at

Y. Ermoliev
International Institute for Systems Analysis (IIASA), Schlossplatz 1, A-2361 Laxenburg, Austria

et al. 2005). The National Research Council (1999) raised the alarm that by 2050 more than a third of the world population will live in seismically and volcanically active zones. Analysis of insurance companies shows that because of economic growth in hazard-prone areas, damages due to natural catastrophes have grown at an average annual rate of 5% (Froot 1997).

The possibility of more frequent catastrophes dominates discussions of current global changes. In fact, one of the main points in the climate change debates concerns possible increase of the frequency of extreme floods, droughts,hurricanes and windstorms (Schiermeier 2006) rather than the increasing global mean temperature which can be within the difference between the average temperature of cities and their surrounding rural areas. Aggravation of catastrophes is associated with increasing interdependencies among different countries. Dantzig (1979) compared our society to a busy highway where disruptions in one of its sections may lead to fundamental traffic jams in other.

The increasing vulnerability of the society calls for new integrated approaches to economic developments and risk management with an explicit emphasis on catastrophes. The standard economic theory is dominated by simplified models of uncertainties and risks, represented by a finite manageable number of contingencies well known to the whole society, which can, therefore, be priced and spread over the whole society through markets. Under such assumptions, catastrophes pose no special problems (Arrow 1996). Insurance risk theory has developed independently of the fundamental economic ideas (see discussion in Arrow 1996; Giarini and Louberg 1978). The central problem of this theory is modeling the probability distribution of total future claims (Grandell 1991), which is then used to evaluate ruin probabilities, premiums, reinsurance arrangements, etc. This theory essentially relies on the assumption of independent, frequent, low-consequence (conventional) risks, such as car accidents, for which decisions on premiums, estimates of claims and likelihood of insolvency (probability of ruin) can be calculated by using rich historical data. The frequent conventional risks also permit simple "more-risks-are-better" strategies with simple "trial-and-error" or "learning-by-doing" procedures for adjusting insurance decisions.

Catastrophes produce losses that are highly mutually dependent in space and time. Therefore, this challenges the standard risk pooling concepts and the standard extremal value theory (Embrechts et al. 2000). The law of large numbers does not apply in this case; the probability of ruin can be reduced not by pooling risks but only if insurers deliberately select the fractions of catastrophic risks they will cover.

The existing extremal value theory deals primarily with independent events and assumes that these events are quantifiable by single numbers. Definitely catastrophes are events not quantifiable in this sense. They may have quite different spatial and temporal patterns, which cause significant dependencies and heterogeneity of losses in space and time. Some of the important dependencies include: the clustering of events in time in a particular region; the spatio-temporal dependencies among climatic events in different regions (such as hurricanes, heavy precipitation, droughts, Al Nino, La Nino, etc.); the dependencies caused by possible cascading events (earthquake → dam failure → flood → technological accident → contamination); the correlation among insurance claims for different policies (such as life, estate, car, employment, business interruption etc.), and at different locations.

While temporal dependencies and characteristics of catastrophes such as, e.g., Guethenberg-Richter law (Amendola et al. 2001; Ermoliev et al. 2000) put into question the use of the traditional Poisson distribution for prediction of extreme events occurrences, the spatial heterogeneities of catastrophes and induced losses emphasize the importance of mitigation and adaptation strategies with proper spatial diversification of the risks. In fact, location-specific losses can be dramatically affected by mitigation decisions (say, by construction of a dike or a flood retention area) and decisions regarding loss spreading schemes within a country or on the international level through the insurance or financial markets.

Many authors stress the need for better models to improve established disaster prevention and mitigation practices (see, for instance, Dilley et al. 2005) and to set guidelines and regulations especially for developing countries. In particular, so-called catastrophe modeling (Walker 1997) is becoming increasingly important for estimating catastrophic losses and making decisions on mitigation measures and/or allocation of coverage, premiums, contingent credits, reinsurance agreements for the residual risks.

This paper shows that, in the presence of catastrophic risks, decisions can be guided by stochastic optimization procedures integrated within catastrophe models. This discussion closely follows the papers by Amendola et al. (2001), Ermoliev et al. (2000) and Ermolieva et al. (2001).

The integrated catastrophe management model reported here can be generalized to account for the interplay between *ex ante* measures, e.g., investment in prevention/mitigation measures (on the part of the public authorities, the citizens and the insurance industry) and ex-post policies for sharing the financial costs after the disaster. Insurance and other financial instruments can be viewed as mitigating catastrophic losses *to a community* by spreading these over a wider region and therefore as decreasing individual financial exposure. Such instruments come into play especially when the costs for further prevention/mitigation measures are prohibitive. The model is therefore useful not only to the insurance industry but also to national authorities in informing decisions on overall catastrophic risk management.

Section 3.2 discusses challenges and difficulties of sound decision-theoretical approaches to insurability of catastrophic risks. Section 3.3 describes a number of case studies stressing the importance of adequate methodological analysis for integrated catastrophe management. Sections 3.4 and 3.5 discuss the integrated catastrophe management model developed at IIASA for a number of catastrophic risks, e.g. floods, earthquakes, windstorms, epidemics, etc. It provides a general framework to bridge decision-oriented economic theory with risk theory and catastrophe modeling. Risk management decisions are evaluated from perspectives of welfare growth in the region. We use economically sound risk measures such as expected costs of overpayments and borrowing, which have strong connection with the insolvency and stability constraints usually assumed in the insurance business and Conditional Value-at-Risk (CVaR) type of risk measures. Section 3.6 discusses the results of a seismic case study for the Tuscany region of Italy. Section 3.7 outlines the computational procedure and Sect. 3.8 summarizes and concludes.

3.2   The Standard Insurance Risk Model

From a formal point of view, insurability of catastrophic risks is equivalent to prevention of certain multidimensional jumping processes to reach vital thresholds provoking insolvency. This is a common problem in risk management.

Consider a simple growth model under shocks, which is a stylized version of an insurance business model (Daykin et al. 1994). The main variable of concern is the risk reserve r^t at time t: $r^t = r_0 + \pi^t - A^t$, $t \geq 0$, where π^t, A^t are aggregated premiums and claims, and r_0 is the initial risk reserve. The process $A^t = \sum_{k=1}^{N(t)} S_k$, where $N(t)$, $t \geq 0$ is a counting process for a number of claims in interval $[0, t]$ (e.g., a Poisson process) with $N(0) = 0$, and $\{S_k\}_1^\infty$ is a sequence of independent and identically distributed random variables (claims), in other words, replicates of a random variable S. The inflow of premiums π^t pushes r^t up, whereas the random outflow A^t pushes r^t down.

The main problem of risk theory (Daykin et al. 1994; Grandell 1991) is the evaluation of the ruin probability $\Psi = P\{r^t \leq 0 \text{ for some } t, t > 0\}$ under different assumptions on π^t and A^t. There are several cases when Ψ can be expressed by an explicit function or at least in a form suited for numerical calculations. An important case arises when the claim distribution is a mixture of exponential distributions and claims occur according to a Poisson process. There are numerous approximations for the probability distribution of A^t. Most of them provide satisfactory results only as long as mean values are concerned and cannot be applied to catastrophes where extreme values are of interest.

The typical actuarial analysis is based on the following. Let us assume that $N(t)$ and S_k are independent, $N(t)$ has intensity α, i.e., $E\{N(t)\} = \alpha t$ and $\pi^t = \pi t$, $\pi > 0$; then the expected profit over the interval $[0, t]$ is $(\pi - \alpha E S)t$; that is, the expected profit increases in time for $\pi - \alpha E S > 0$. The difference $\lambda = \pi - \alpha E S$ is the "safety loading". The law of large numbers implies that $\lfloor \pi^t - A^t \rfloor / t \rightarrow [\pi - \alpha E S]$ with probability 1. Therefore, in the case of positive safety loading $\pi > \alpha E S$ we have to expect that the real random profit $\pi^t - A^t$ would also be positive for large enough t under the appropriate choice of premium $\pi = (1 + \rho)\alpha E S$, where ρ is the "relative safety" loading, $\rho = (\pi - \alpha E S)/\alpha E S$. But this holds only if the ruin does not occur before time t. This is a basic actuarial principle: premiums are calculated by relying on the mean value of aggregated claims increased by the (relative) safety loading. Thus, practical actuarial approaches ignore complex interdependencies among timing of claims, their sizes, and the possibility of ruin, $r^t \leq 0$. The random jumping process r^t is often simply replaced by a (linear in t) function $\bar{r}^t = r_0 + (\pi - \alpha E S)t$. In the case of positive expected profit, $\pi - \alpha E S > 0$, the expected risk reserve $\bar{r}^t$

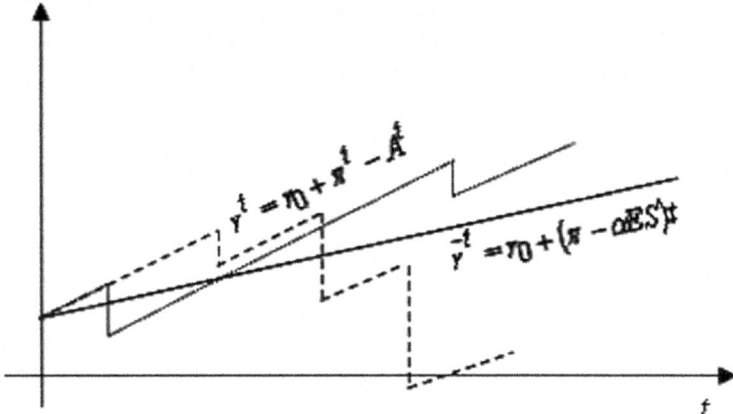

Fig. 3.1 Stochastic trajectory of the risk reserve

increases linearly in t, as shown in Fig. 3.1. This is a basic actuarial principle: premiums are calculated from the mean value of aggregate claims A^t increased by a safety loading $\lambda > 0$, $\pi = (1 + \lambda)\alpha ES$. As we can see from Fig. 3.1, although $\bar{r}^t$ increases, insolvency may occur. It depends on the existence of large claims. To avoid insolvencies, λ must be chosen properly. The estimation of ES becomes an extremely complicated task in the case of rare catastrophic events with relatively small historical data. The law of large numbers does not operate, since catastrophes produce highly dependent losses and claims. "Learning-by-doing" or "what-if" scenario analysis approaches may be very expensive, dangerous and even simply impossible. Instead, the role of catastrophe modeling and stochastic optimization techniques becomes essential for making decisions on risk sharing and mitigation measures. Catastrophic losses and claims at different geographical locations and their dependencies on policy variables can be simulated with a help of catastrophe models. Stochastic optimization makes it possible to adjust decision variables towards desired policy options in response to simulated catastrophic events and available historical data.

In general, various decision variables affect the probability of insolvency Ψ. Claim size S depends on the coverage of the insurer in different locations. Important decision variables are r_0, π, and reinsurance arrangements, for example, the "excess of loss" reinsurance contract. If the latter is established, the insurer retains only a portion, $S(x) = \min\{S, x\}, x \geq 0$, of a claim S and the remaining portion is passed to the reinsurer. The reinsurance contracts with deductibles are defined by two variables $x = (x_1, x_2)$. In this case $S(x) = \max\{x_1, \min[S, x_2]\} - x_1, x_1 \geq 0, x_2 \geq 0$ is retained by the insurer. The reduction of Ψ to acceptable levels can be viewed as a chance constraint problem (Ermoliev and Wets 1988). The complexity is associated with the jumping process A^t and with analytically intractable dependencies of A^t on decision variables, which restricts the straightforward use of conventional stochastic optimization (STO) methods. The direct sample mean estimation of $\Psi(x)$ requires a very large number of observations and it leads to discontinuous functions.

To overcome these difficulties, the following simple idea can be used for rather general problems.

Consider $t = 0, 1, \ldots$ and assume that r^t can be subdivided into a "normal" part (including r_0), M^t, associated with ordinary claims, and a "catastrophic" part B^t, $\pi^t = \pi t$, where π is the rate of premiums related to catastrophes; the probability of a catastrophic event p is characterized by a probability distribution in an interval $\left[\underline{p}, \overline{p}\right]$, and the probability distribution $V_t(z) = P[M^t < z]$ can be evaluated. Assume also that ruin may only occur due to a catastrophe. Then the probability of ruin after the first catastrophe and with the "excess of loss" contract is defined as a function

$$\Psi(x) = E \sum_{t=1}^{\infty} p(1-p)^{t-1} V_t(\min\{x, B^t\} - \pi t). \tag{3.1}$$

The search for a desirable x can be based on methods outlined in Sect. 3.7.

3.3 Overview of Case Studies

The model (3.1) has a rather simplified illustrative character. In reality, damages and claims depend on geographical patterns of catastrophes, clustering of property values in the region, available mitigation measures and regulations, and the spread of insurance coverage among different locations. For all these reasons, the model should be geographically explicit (Ermoliev et al. 2000) for the description of property values and insurance contracts in different parts of the region, as well as for modeling of catastrophes.

Catastrophic modeling (see Walker 1997) is important to insurance companies for making decisions on the allocation and values of contracts, premiums, reinsurance arrangements and mitigation measures. Using these models it is possible to simulate different patterns of catastrophes as they may happen in reality and analyze their impacts. Different catastrophic scenarios lead, in general, to different "optimal" decision strategies. The important question is how we can find a decision strategy, which is the "best" against all possible scenarios of catastrophes. Ermoliev et al. (2000) show that the search for such decisions can be done by incorporating stochastic optimization techniques into catastrophe modeling. Indeed by this approach, it is possible to take into account complex interdependencies between damages at different locations, available decisions and resulting losses and claims. To better understand the main features of the model described in the next section, let us sketch out some case studies where the model has been applied (Amendola et al. 2000a, b; Ermolieva et al. 2001).

The main concern of these case studies is related to issues emphasized by Froot (1997): "most of the catastrophic losses are paid ex-post by some combination of insurers and reinsurers (and their investors), insured, state and federal agencies and taxpayers, with only some of these payments being explicitly arranged ex-ante". This introduces considerable uncertainty about burden sharing into the system, with no particular presumption that the outcome will be fair. The result is incentives for players to shift burdens towards others, from the homeowner who builds on exposed coastline, to insurers who write risks that appear highly profitable in the absence of a large event. In Hungary, given its problems of a poor and immobile population, ex-ante mechanisms to fund the costs of recovery and, in particular, the establishment of a multipillar flood loss-sharing program, are especially important. In the analysis of the Upper Tisza river pilot region (Ermolieva et al. 2001) it is assumed, in particular, that for the first pillar the government would provide compensation of a limited amount to all households that suffer losses from flooding. As the second pillar, a special regional fund would be established through a mandatory public flood insurance on the basis of location-specific risk exposures. It is assumed that the governmental financial aid is regulated through this fund. As a third pillar, a contingent credit may also be available to provide an additional injection of capital to stabilize the system. In the latter case, the lender charges a fee that the borrower (in our case, the fund) pays as long as the trigger event does not occur. If the event does occur, the borrower may rapidly receive a bond or a contingent credit (see e.g. Ermolieva et al. 2001). Such a program would increase the responsibility of individuals and local governments for flood risks and losses. Local governments may be more effective in the evaluation and enforcement of loss-reduction and loss-spreading measures, but this is possible only through location-specific analysis of potential losses, the mutual interdependencies of these losses, and the sensitivities of the losses to new risk management strategies.

The analysis of possible gains and losses from different arrangements of the program outlined above is a multi-disciplinary task, which has to take into account the frequency and intensity of hazards, the stock of capital at risk, its structural characteristics, and different measures (in particular, engineering, financial) of vulnerability. For this purpose Ermolieva et al. (2001) discusses a GIS-based catastrophe model developed for the Upper Tisza pilot region that, in the absence of historical data, simulates samples of mutually dependent potential losses at different locations. The model emphasizes the cooperation of various agents in dealing with catastrophes. The solution to catastrophic risk management, especially for small economies with limited risk absorption capacity, cannot be accomplished (see Pollner 2000; Amendola et al. 2000a, b; Cummins and Doherty 1996) without pooling of risk exposures. The proposed model involves pooling risks through mandatory flood insurance based on location-specific exposures, partial compensation to the flood victims by the central government, and a contingent credit to the pool. Definitely, this program encourages accumulation of own regional capital to better "buffer" international reinsurance market volatility. In order to stabilize the program such economically sound risk indicators as expected overpayments by "individuals" and an expected shortfall of the mandatory insurance are used. The analysis is oriented towards the most destructive scenarios. It was shown (see Ermoliev et al. 2000, 2001) that the explicit introduction of ex-post borrowing (see also Sect. 3.5) is a valid measure against insolvency.

For the seismic prone Irkutsk region in Russia (Amendola et al. 2000a, b) the focus of the analysis was on the feasibility of an insurance pool to cover catastrophic losses subject to strong standard insolvency regulations. In contrast to the case study in Hungary, the contribution of different insurers to the stability of the pool was explicitly analyzed by taking into account the transaction costs and effects of mutual dependencies among claims from different locations.

In Russia (Amendola et al. 2000a, b), new legislative instruments and government resolutions are creating a framework for risk management similar to that existing in the OECD countries. However, in Russia and other transition countries the emergence of a viable insurance industry is slow and subject to insolvency risks due to problems of the national economies, the lack of consolidated experience and practicable guidance, and the lack of sufficient risk reserves of the existing companies. For example, when in seismic regions insurance is available, premiums are neither based on the probability of occurrence of earthquakes nor do they differentiate among geological situations and construction type. The model proposed in Amendola et al. (2000a, b, 2001) is a pilot exercise, which, however, can create the basis for cooperation among researches, insurers, and regulatory bodies in transition countries. In the case of Russian emerging insurance industry, cooperation among insurers will undoubtedly play an important role in stabilizing the insurance market. A key problem, however, is the lack of necessary information on the distribution of losses among locations. For this purpose a region-specific earthquake generator (see Baranov et al. 2002; Rozzenberg et al. 2001) was designed and incorporated within the STO model (Amendola et al. 2000a, b).

In most cases neither the market nor the government alone may be acceptable as the mechanism for catastrophic risk management. Thus, some form of a public-private partnership may be appropriate (Kunreuther and Roth 1998).

A well-known example of a government acting as a primary insurer is the U.S. National Flood Insurance Program (NFIP), which seeks to provide insurance at actuarially fair premiums combined with incentives for communities and homeowners to take appropriate loss-reducing measures. Given the size of the U.S. and the large number of persons living in flood plains, the program is sufficiently diversified to cover most regional losses with premium payments. In contrast to the NFIP, some government insurance schemes in Europe, e.g., the French national insurance program, cross-subsidies claims. This is because the Constitution (1946, 1958) established the principle of "the solidarity and equality of all French citizens facing the expenses incurred through national calamities" (Gilber and Gouy 1998).

However, even if many governments are pursuing policies to reduce their role in compensating victims, a study (Linnerooth-Bayer and Amendola 2000) confirms that the victims and their governments bear the major losses from natural disasters and, world-wide, there is only moderate risk-transfer with insurance. An important consideration for national insurance strategies is linking private insurance with mitigation measures to reduce losses. Insurers, however, are reluctant to enter markets that expose them to a risk of bankruptcy. In the U.S., for example, many insurers pulled out of catastrophic risk markets in response to their large losses from natural catastrophes in the last decade (Insurance Service Office 1994).

To reduce their risk of insolvency, insurers' strategies may be based on modelling tools that account for the complexity implied by the manifold dependencies in the stochastic process of catastrophic events, decisions and losses. For example, to study the problem in its complexity for the Tuscany region, a spatial-dynamic, stochastic optimisation model was developed in Ermolieva et al. (1997), Ermoliev et al. (2000, 2001), which is described below.

In Italy, a law for integrating insurance in the overall risk management process was proposed only in late 1997 (within the Design of Law 2793: "Measures for the stabilisation of the public finance"). This opened a debate, which has not yet been concluded by a legislative act. Therefore policy options for a national insurance strategy are still open to investigation. The Institute for Research on Seismic Risk of the Italian National Research Council made data from a previous study available (Petrini 1995). These have been incorporated in a Monte Carlo catastrophe model, which simulates occurrence of earthquakes affecting the region, calculates attenuation according to the geological characteristics, and finally determines the acceleration at the ground in each municipality. The model explicitly incorporates the vulnerability of the built environment, with data on number and types of buildings in each municipality of the region.

3.4 Stochastic Optimization (STO) Model

Catastrophes may lead to large costs, social disruption and economic stagnation. A catastrophe would ruin many agents if their risk exposures are not properly managed. To design safe catastrophic risk management strategies it is necessary to define at least the following: patterns of possible disasters in space and time, a map of regional values and their vulnerability, and feasible decisions, e.g., insurance coverages. The model of this section uses this information. It emphasizes the collective nature of catastrophe risk management. The aim of this model is to address only main features of the problem. Basically, we assume that the goal of the insurance is to maximize its wealth through gathering of capital reserve while maintaining survival and stability (see also the discussion in Stone 1973). In a similar manner, other agents are concerned with their sustained wealth growth. The model emphasizes catastrophe risk management as a long-term business rather than as subject of annual accounting and taxation. Accordingly, catastrophe reserves should be accumulated over years.

Assume that the study region is divided into sub regions or cells $j = \overline{1, m}$. A cell may correspond to a collection of households, a zone with similar seismic activity, a watershed, a grid with a segment of a gas pipeline, etc. The choice of cells provides a desirable representation of losses. For each cell j there exists an estimate of its "wealth" at time t that may include the value of infrastructure, houses, factories, etc. A sequence of random catastrophic events $\omega = \{\omega_t, \ t = \overline{0, T-1}\}$ affects different cells $j = \overline{1, m}$ and generates at each $t = \overline{0, T-1}$ mutually dependent losses $L_j^t(\omega)$, i.e., damages of the wealth at j, T is a time horizon. These losses can be modified by various decision variables. Some of the decisions reduce losses, say, a dike in the case of flood risks, whereas others spread them on a regional, national, and international level, e.g., insurance contracts, catastrophe securities, credits, and financial aid. If x is the vector of the decision variables, then the losses $L_j^t(\omega)$ are transformed into $L_j^t(x, \omega)$. For example, we can think of $L_j^t(x, \omega)$ as $L_j^t(\omega)$ being affected by the decisions of the insurance to cover losses from a layer $[x_{j1}, x_{j2}]$ at a cell j in the case of a disaster at time t:

$$L_j^t(x, \omega) = L_j^t(\omega) - \max\left\{x_{j1}, \min\left[x_{j2}, L_j^t\right]\right\} + x_{j1} + \pi_j^t$$

where $\max\left\{x_{j1}, \min\left[x_{j2}, L_j^t\right]\right\} - x_{j1}$ are retained by insurance losses, and π_j^t is the premium.

In the most general case, the vector x comprises decision variables of different agents, including governmental decisions, such as the height of a new dike or a public compensation scheme defined by a fraction of total losses $\sum_{j=1}^m L_j^t$. The insurance decisions concern premiums paid by individuals and the payments of claims in the case of catastrophe. There are complex interdependencies among these decisions, which call for the cooperation of agents. For example, the partial compensation of catastrophe losses by the government enforces decisions on loss reductions by individuals and, hence, increases the insurability of risks, and helps the insurance to avoid insolvency. On the other hand, the insurance combined with risk-reduction measures can reduce losses, compensations and governmental debt and stabilize the economic growth of the region and the wealth of individuals.

We assume that ω is an element of a probability space $(\Omega, \mathbf{F}, P)$, where Ω is a set of all possible ω, and $\mathbf{F}$ is a $\boldsymbol{\sigma}$-algebra of measurable (with respect to probability measure P events from Ω. Let $\{\mathbf{F}_t\}$ be a non-decreasing family of $\boldsymbol{\sigma}$ algebras, $\mathbf{F}_t \subseteq \mathbf{F}_{t+1}, \mathbf{F}_t \subseteq \mathbf{F}$. Random losses $L_j^t(\omega)$ are assumed to be F_t – measurable, i.e., they depend on the observable catastrophes until time t. In the following we specify dependencies of these variables on ω, although sometimes we do not use ω when these dependencies are clear from the text.

Catastrophe losses are shared by many participants, such as individuals (cells), governments, insurers, reinsurers, and investors. In the model we call them "agents", since the main balance equations of our model are similar for all of them. For each agent i a variable of concern is the wealth W_i^t at time $t = \overline{0, T}$

$$W_i^{t+1}(\omega) = W_i^t(x, \omega) + I_i^t(x, \omega) - O_i^t(x, \omega), \; i = \overline{1, n}, \; t = \overline{0, T-1}, \; \omega \in \Omega \quad (3.2)$$

where W_i^0 is the initial wealth. This is a rather general process of accumulation, which, depending on the interpretation, can describe the accumulation of reserve funds, the dynamics of environment contamination process, or processes of economic growth with random disturbances (shocks), reserves of the insurance company at moment t, the gross national product of a country or the accumulated wealth of a specific region. In more general cases, when catastrophes may have profound effects on economic growth, this model can be generalized to an appropriate version of an economic-demographic model (see, for example, MacKellar and Ermolieva 1999) enabling one to represent movements of individuals and the capital accumulation processes within the economy.

For the simplicity of the exposition we do not discuss discount rates in these equations since catastrophes require non-standard approaches. In particular, induced by catastrophes, discount rates become important, which is evident from the evaluation (3.1). We use also the same index i for quite different agents. Therefore, the variables $I_i^t(x, \omega)$, $O_i^t(x, \omega)$ may have quite a different meaning. For example, for each insurer i we can think of I_i^t as premiums π_i^t which are ex-ante arranged and do not depend on ω, whereas O_i^t is defined by the claim size S_i^t and possible transaction costs which triggers a random jump of the risk reserve W_i^t (usually denoted as R_i^t) downwards at random times of catastrophic events (as in the simple model of Sect. 3.2). If i corresponds to a cell, then income I_i^t may be affected by a catastrophic event ω generated by a catastrophe model. The incomes I_i^t can be defined by a set of scenarios or through a regional growth model with geographically explicit distribution of the capital among cells. The term O_i^t may include losses L_i^t, taxes and premiums paid by i. For central or local governmental agent i (e.g., mandatory insurance, catastrophe fund) I_i^t may include a portion of taxes collected by the government (compensations of losses by the government), and O_i^t may consist of mitigation costs, debts, loans and fees paid for ex-ante contingent credits.

Catastrophes may cause strong dependencies among claims S_i^t for different insurers i. These claims are defined by decisions on coverages of losses L_j^t from different locations j. For example, let us denote by x_{ij}^t a searched fraction of L_j^t covered by insurer i, e.g., assume $i = \overline{1, n}$. Then

$$\sum_{i=1}^{n} x_{ij}^t \leq 1, x_{ij}^t \geq 0, \ j = \overline{1, m} \tag{3.3}$$

and claims S_i^t are linear functions of

$$x = \left\{ x_{ij}^t, \ i = \overline{1, n}, \ j = \overline{1, m}, \ t = \overline{0, T-1} \right\} :$$

$$S_i^t(x, \omega) = \sum_{j=1}^{m} L_j^t x_{ij}^t, i = \overline{1, n}, t = \overline{0, T-1}.$$

If I_i^t, O_i^t simply correspond to premiums π_i^t and claims S_i^t, then the wealth of insurer i (its risk reserves) are calculated for $t = \overline{0, T-1}$, $\omega \in \Omega$ as follows

$$R_i^{t+1}(x, \omega) = R_i^t(x, \omega) + \sum_{j=1}^{m} \pi_{ij}^t x_{ij}^t - \sum_{j=1}^{m} L_j^t(\omega) x_{ij}^t \tag{3.4}$$

where π_{ij}^t are rates of premiums per unit of coverage.

For each i consider a stopping time τ_i for process $W_i^t(x, \omega)$, i.e., a random variable with integer values, $t = \overline{0,T}$. The event $\{\omega : \tau_i = t\}$ with fixed t depends only on the history till t and it corresponds to the decision to stop process $W_i^t(x, \omega)$ after time t. Therefore, τ_i in the case of W_i^t defined according to (3.4) depends on $\left\{ x_{ij}^k, \ i = \overline{1,n}, \ j = \overline{1,m}, \ k = \overline{0,t} \right\}$, i.e., it is a function $\tau_i(x, \omega)$. Examples of τ_i may be $\tau_i = T$, the time of the first catastrophe, or the time of the ruin before a given time T: $\tau_i(x, \omega) = \min\{T, \min\lfloor t : W_i^t(x, \omega) < 0, \ t > 0 \rfloor\}$. The last example defines τ_i as a rather complex implicit function of x.

Assume that each agent i maximizes (possibly negative) "wealth" at $t = \tau_i$. The notion of wealth at t requires exact definition since it must represent, in a sense, the whole probability distribution W_i^t. The traditional expected value EW_i^t may not be appropriate for probability distributions of W_i^t affected by rare catastrophes of high consequences. As a result they may have a multimode structure with "heavy tails". We can think of the estimate for W_i^t as a maximal value V_i^t, which does not overestimate, in a sense, random value W_i^t, i.e., cases when $\min_{s \leq t} \left(W_i^s(q, \omega) - V_i^t \right)$ <0. Formally, V_i can be chosen by maximizing

$$V + \gamma E \min\{0, W_i^t - V\} \tag{3.5}$$

or more general function $V + \gamma Ed\left(W_i^t - V\right)$, for appropriate function $d(\cdot)$ and $\gamma > 0$. The second term in (3.5) can be considered as the risk of overestimating of the wealth $W_i^s(x, \omega)$ for $s = 0, 1, \ldots, t$. This concept corresponds to the CVaR risk measure (see Artzner et al. 1999; Jobst and Zenios 2001; Rockafellar and Uryasev 2000). The maximization of (3.5) is a simple example of the so-called stochastic maximin problems. It is easy to see from the optimality conditions for this problem (see Ermoliev and Wets 1988, pp. 165, 416, and further references) that for continuous distributions the optimal value V satisfies condition $P\lfloor W_i^t \leq V \rfloor = 1/\gamma$. For the normal distribution and $\gamma = 2$, it coincides with the traditional mean value EW_i^t. In the case of quadratic function $d(\cdot)$ and $\gamma = \infty$, i.e., the maximization of $E\left(W_i^t - V\right)^2$, the optimal $V = EW_i^t$.

Besides the maximization of wealth, the agent i is concerned with the risk of insolvency, i.e., when $W_i^s < 0$ for some $s = 0, 1, \ldots, t$ as well as the lack of sustained growth, i.e., when $I_i^s - O_i^s < 0$ for some $s = 0, 1, \ldots, t$. In accordance with this consider the stochastic goal functions

$$f_i^t(x, V, \omega) = V_i^t + \gamma_i \min\left\{ 0, \min_{s \leq t} \left[W_i^s(x, \omega) - V_i^s \right] \right\} +$$

$$\delta_i \min\left\{ 0, \min_{s \leq t} W_i^s(x, \omega) \right\} + \beta_i \min\left\{ 0, \min_{s \leq t} \left[I_i^s(x, \omega) - O_i^s(x, \omega) \right] \right\}$$

$$F_i(x, V) = E f_i^{\tau_i(x, \omega)}(x, V, \omega), \tag{3.6}$$

where nonnegative γ_i, δ_i, β_i are substitution coefficients between wealth V_i^t and risks of overestimating wealth, insolvency, and overestimating sustained growth.

If a catastrophe is considered as the most destructive event, then we can use in the definition of f_i^t simply $s = t$ instead of $\min_{s \leq t}$. These requirements reflect survival and stability constraints of agents. In (3.6) we use a modified form of (3.5), which is more appropriate for dynamic problems. Each agent attempts to maximize $F_i(x, V)$.

Pareto optimal improvements of risk situations with respect to goal functions F_i (x, V) of different agents can be achieved by maximizing

$$W(x, V) = \sum_{i=1}^{n} \alpha_i F_i(x, V) \tag{3.7}$$

for different weights $\alpha_i \geq 0$, $\sum_{i=1}^{n} \alpha_i = 1$. These weights reflect the importance of the agents. The maximization of $W(x, V)$ for different weights α_i, $i = \overline{1, n}$, corresponds to a stochastic version of the welfare analysis (Ginsburg and Keyzer 1997).

When $n > 1$ this model generalizes Borch's (1962) fundamental ideas of risk sharing to the case of catastrophic risks. In the Borch model risks from different locations are substitutable, and the insurance pool is concerned only with the redistribution of the total "risk mass". According to (3.3), our model emphasizes differences among risks from different locations, i.e., $m > 1$ in contrast to $m = 1$ of Borch's model.

Random functions $f_i^t(x, V, \omega)$ have a complex nested analytically intractable structure defined by simulated patterns of catastrophes. Their non-smooth character is due to the presence of operators min and stopping times τ_i, which may be complex implicit functions of (x, ω). When (4) $f_i^t(x, V, \omega)$ are concave functions in x as min of linear functions. Hence, expectations $F_i^t(x) = E f_i^t(x, V, \omega)$ are also concave functions in x for fixed t. The use of stopping times, $t = \tau_i$, generally destroys their concavity and even continuity. If stopping times do not depend on x, then these expectations are also concave. The use of such risk functions as in (3.5) is similar to the Markowitz (1987) mean-semivariance model and the Konno and Yamazaki (1991) model with absolute deviations. Connections of problems (3.5) with the CVaR risk measure are established in Rockafellar and Uryasev (2000).

The choice of weights (risk coefficients) γ_i, δ_i, β_i, provides different trade-offs between wealth and risks. The increase of these parameters better eliminates corresponding risks.

3.5   Insolvency, Stopping Time, and Nonsmooth Risk Functions

A key issue for selecting catastrophic risk portfolios is the financial ruin of insurers. It was shown (see Ermoliev et al. 2000), that when risk coefficients γ_i, δ_i, β_i in (3.7) become large enough, then the probability of associated risks, in particular the probability of ruin, drops below a given level p:

$$P\left[\min_{s\le\tau_i} W_i^s < 0, \ i = \overline{1,n}\right] \le p. \tag{3.8}$$

The maximization problem defined by (3.6) and (3.7) is much simpler than the problem defined in terms of the chance constraint (3.8). The functions $F_i(x)$ defined according to (3.4) for $W_i^t(x,\omega) := R_i^t(x,\omega)$ are concave, whereas constraints (3.8) for the same case may have discontinuities character, e.g., if ω has a discrete distribution. The problem defined in terms of the chance constraints (3.8) has a convex feasible set only under a strong assumption on the probability measure.

The discontinuous nature of the problem (3.6) and (3.7) may still be connected with the stopping time defined as the ruin (insolvency) moment. Different smoothing techniques for this case are analyzed in Ermoliev et al. (2001). In particular, a rather natural idea of smoothing consists of introducing the possibility of borrowing money in the case of insolvency. It is natural to expect that when the payment for borrowing is high, agents will tend to exclude such a necessity through a reasonable selection of the portfolios, i.e., to keep constraints on the insolvency within reasonable limits. Let us slightly modify the process (3.2):

$$W^{t+1}(x,y,\omega) = W^t(x,y,\omega) + I^t(x,\omega) - O^t(x,\omega) + y_{t+1} - (1+\beta_t)y_t \tag{3.9}$$

where for the simplicity of notation we do not use here index i, y_t is a value of borrowing on the interval $[t-1,t)$, β_t is the bank interest for the credit on the interval $[t-1,t)$, and $y = \{y_0,\dots,y_T\}$. According to (3.9), the borrowing taken out at the moment t to maintain solvency should be paid off at the next instant of time $t+1$ with interest β_t. If the reserves of the company are not sufficient for this purpose, then new borrowings are taken. The following fact is the key for dealing with discontinuities of the stopping time effects and the insolvency constraints. Let us represent the process $W^t(x,y,\omega)$ as

$$W^t(x,y,\omega) = \overline{W}^t(x,\omega) - \sum_{s=1}^{t-1}\beta_s y_s + y_t, \ \overline{W}^t(x,\omega)$$

$$= W^0 + \sum_{s=1}^{t}(I^s(x,\omega) - O^s(x,\omega))$$

and let $(x^*(\beta), V^*(\beta), y^*(\beta))$ be a solution of the following problem: maximize

$$F(x,V) = E\max_{y\ge0}\lfloor f^T(x,V,y,\omega) - (1+\beta_T)y_T\rfloor, \ W^t(x,y,\omega) \ge 0, \ 0\le t\le T \tag{3.10}$$

where $f^t(x,V,y,\omega)$ is defined as in (3.6) for $W^t(x,y,\omega)$ defined according to (3.9).

Theorem. (Ermoliev et al. 2001): Assume that $\overline{R}_t(0) \geq 0$, $P\left[\overline{W}^t(x, \omega) = 0\right] = 0$,

for any $x \in X$, $t = \overline{1,T}$. Then the probability of borrowing can be arbitrary small by taking interest coefficients β_t, $t = \overline{1,T}$, large enough, i.e., $P\lfloor\overline{W}^t(x^*(\beta_t), \omega) \geq 0$, $t = \overline{1,T}\rfloor \to 1$ a.s. for $\min_{1 \leq t \leq T} \beta_t \to \infty$.

3.6   The Tuscany Region Case Study

In this section we specify the general model described in Sect. 3.4 to the Tuscany region, Italy (Amendola et al. 2000a, b). Even if Tuscany is not among the most hazardous regions with respect to seismic activities, the case study is quite representative for the methodological approach. The choice of the region was determined by the fact that the Institute for Research on Seismic Risk of the Italian National Research Council made models and data from a previous study available (Petrini 1995). These were incorporated in a Monte Carlo generator of seismic events, which simulates occurrence of earthquakes affecting the region, calculates attenuation according to the geological characteristics, and finally determines the acceleration at the ground in each municipality. The IIASA spatial-dynamic, stochastic optimisation model has been customised to explicitly incorporate the vulnerability of the built environment, with data on number and types of buildings in each municipality of the region. The study focused on the analysis of different policy options for an insurance program and on the interplay between investments in physical mitigation (retrofitting) and risk-sharing measures should be investigated.

The region was subdivided into $M \approx 300$ sub-regions, which corresponds to the number of its municipalities. For each municipality j, number and types of buildings, their vulnerability, and number of built cubic meters are available. These represent the so-called estimate of "wealth" W_j in the municipality j. Using data and models in Petrini (1995), a catastrophe generator was created (see Amendola et al. 2000a, b; Baranov et al. 2002; Rozzenberg et al. 2001) using Gütenberg – Richter law and the attenuation characteristics of the region (see Fig. 3.2). This enables one to generate the occurrences of earthquakes at random times, including their intensities and ground accelerations in each municipality. The generator could be easily adapted to incorporate different kinds of hazard distributions, non-poissonian catastrophic processes, as well as micro-zoning within a municipality. It produces earthquake scenarios at random time moments according to geo-physical characteristics of faults and soil type.

Simulated in time and space, earthquakes $\omega_0, \ldots, \omega_t$ may occur at different municipalities, inside or outside the region, have random magnitudes and, therefore, affect a random number of municipalities.

In municipalities affected at time t the vulnerability relations between accelerations and losses (Petrini 1995) according to the type (masonry or reinforced concrete), age and maintenance of the buildings are used to estimate the number of cubic meters of destroyed properties. The economic loss of destroyed cubic meters

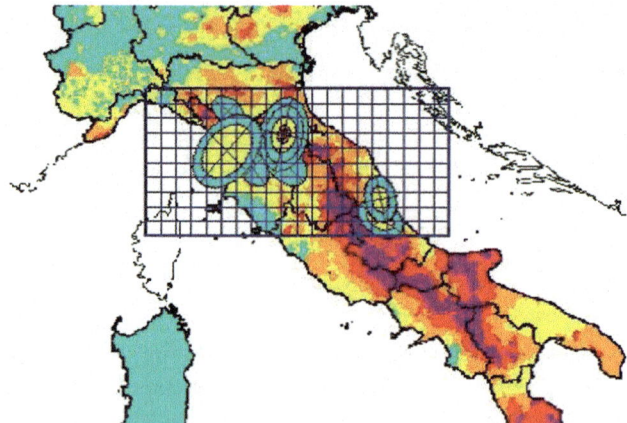

Fig. 3.2 Earthquake generator

of a building is defined as the cost for their reconstruction. Then it is possible to be independent of contingent pricing by considering the cost of reconstruction per cubic meters to be the monetary unit. In this way the simulation of time histories for possible earthquakes in the region produces the sets of economical losses, and enables the design of an insurance programme. It also enables one to determine in which way preventive retrofitting could decrease the losses: this is easily done by a consequent decrease of the vulnerability indices in the loss model. In this way it is possible to study the interplay between structural measures and risk-sharing for an integrated risk management approach, and to design an insurance system linked to incentives for retrofitting of the built environment. Our analysis was primarily concerned with the following. In its early version the Italian Design of Law 2793 (1998) to reduce the impact of natural disasters on the governmental budget, included provisions for an insurance programme against all natural hazards. It was not intended to make this insurance mandatory, but to make mandatory the extension of a fire insurance policy to all natural hazards, in a way similar to the French system (see Sect. 3.3). In addition to tax incentives for such an insurance, it stipulated a maximum exclusion layer of 25%, the creation of a pool of insurance companies with an appropriate reserve fund, e.g., corresponding to the annual average government payment for compensating losses (with some forms of state guarantee to be specified further), and linking of the premium to the premium for fire policy. This article was withdrawn, and later proposals are still subject of discussion.

Starting from these principles, the case study intends to demonstrate how the model evaluates and offers the decision-makers different policy options. Let us assume that an insurance company (this might be a pool of companies or the government itself acting as an insurer) covers a fraction, e.g., $q = 0.75$, of earthquake losses. The rest $v = 1 - q$, according to the Design of Law, would be compensated by the state. The state would also required to contribute to the reserve funds in case of excessive losses.

The company has an initial catastrophe fund or a risk reserve R^0, which in general is characterized by a random variable dependent on past catastrophic events. It is also possible to analyze necessary for the future as R^0 a policy variable. For example, taking $R^0 = 0$ enables us to evaluate the capacity of the region to accumulate risk reserves in the future. Assume that the time span consists of $t = \overline{1,T}$, $T = 50$, time intervals. The stopping time τ is the time of the first catastrophe in the region within the time horizon T. The risk reserve (wealth) R^t of the pool at time $t = \overline{1,T}$, is calculated according to (3.4):

$$R^t = R^{t-1} + \sum_{j=1}^{m} \pi_j - \sum_{j=1}^{m} L_j^t(\omega_t)q$$

where q defines the coverage of the pool in affected municipalities j at time t, π_j is the premium rate from the municipality j, $L_j^t(\omega_t)$ is the loss (damage) at j caused by the simulated catastrophic event ω_t at time t. The value $L_j^t(\omega_t)$ depends on the event ω_t, the content of j, mitigation measures and deterioration of the built environment. The analytical structure of the probability distribution of the random variable R^t is intractable, therefore, the methodology relies on Monte Carlo simulation.

Standard actuarial approaches calculate premiums in accordance with loss expectations. Therefore this study analysed two policy options based on similar principles:

1. Premiums based on the average damage over all municipalities (solidarity principle, bringing less exposed locations to pay premiums equal to more severely exposed ones, as in the spirit of the proposed insurance programme)
2. Location-specific premiums based on average damage in the particular municipality, i.e., risk-based premiums.

   However, the use of average losses may be misleading in the case of heavy tailed distributions which are typical for catastrophic losses. The stochastic optimisation allows the analysis of different criteria and takes into account dependencies among location specific losses. As an important example, a third policy option has been considered:
3. Premiums calculated in a way that equalises in a fair manner the risk of instability for the insurance company and the risk of premium overpayment for exposed municipalities. Besides this, it was important to analyse location specific coverages and the amount of governmental compensation as a decision variable.

For *Option 3* it was assumed that the pool maximises its wealth (risk reserves) taking into account the risks of the insolvency under the constraint on "fair" premiums. "Fair" premiums are defined according to the specified probability (say, once in every 100 years) of cases when paid premiums exceed actual claim sizes.

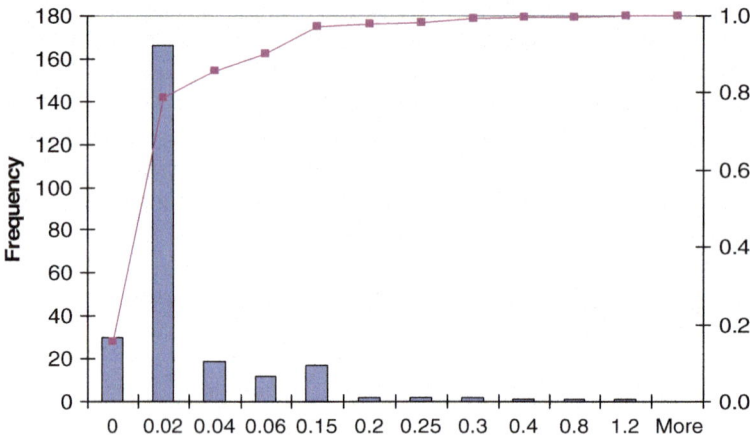

Fig. 3.3 Distribution of municipality-specific premiums (per building volume/municipality)

Accordingly, the goal function (3.6) for the pool at $t = \tau$, $R^0 = 0$, is

$$f^{\tau}(x, V, \omega) = V + \gamma \, \min\{0, R^{\tau}(x, \omega) - V\} + \delta \, \min\{0, R^{\tau}(x, \omega)\}.$$

The stability of the welfare growth of municipalities can be written in the form of the chance constraints on overpayments:

$$P\left\{(1 - q)L_j^t + L_j^t q_j < \pi_j\right\} \leq p, \; \sum_{j=1}^{m} q_j = q,$$

where $x = (\pi_1, \ldots, \pi_n, q_1, \ldots, q_n), x \geq 0, p$ is a given "safety" level. The difference $q - q_j$ defines the partial coverages of some municipalities, which generates the demand for further increase of the compensation by the government. The wealth of municipalities at $t = \tau$ changes due to the insurance program from $W_j^{\tau} - L_j^{\tau}$ to $W_j^{\tau+} = W_j^{\tau} - L_j^{\tau} + (1 - q + q_j)L_j^{\tau} - \pi_j$. The stochastic goal function (3.6) for municipality j at $t = \tau$ is defined as

$$f_j^{\tau}(x, V_j, \omega) = V_j + \gamma_j \, \min\left\{0, W_j^{\tau+} - V_j\right\} +$$
$$\delta_j \, \min\left\{0, (1 - q + q_j)L_j^{\tau} - \pi_j\right\}$$

Figures 3.3, 3.4, 3.5, 3.6, and 3.7 illustrate some numerical results. The number of simulations is shown on the vertical axis.

For *Option 1*, where the burden of losses is equally distributed over the population, the simulation of catastrophic losses showed that the annual premium is equal to the flat rate of 0.02 monetary units (m. u.) per cubic meter of building.

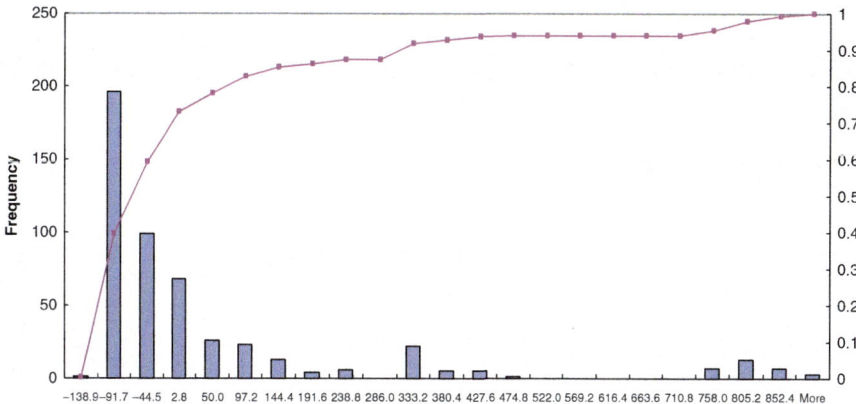

Fig. 3.4 Distribution of insurer's reserve, *Options 1, 2* (thousands m.u., 50 years)

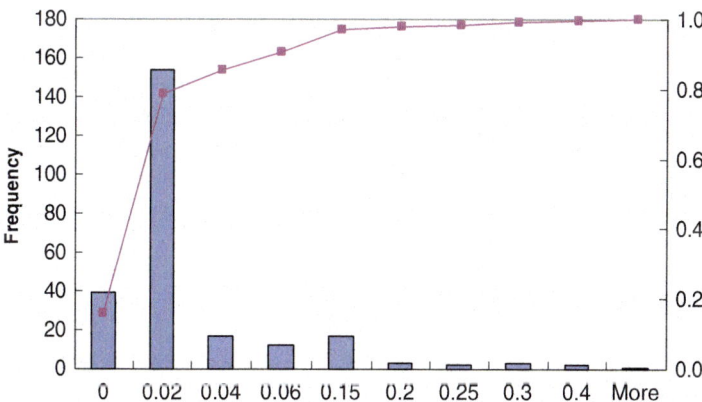

Fig. 3.5 Distribution of "Fair" premiums, *Option 3*, (per building volume/municipality)

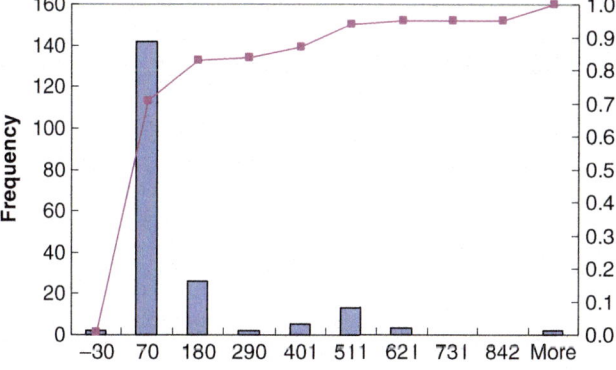

Fig. 3.6 Distribution of insurers' reserve, *Option 3* (thousands m.u., over 50 years)

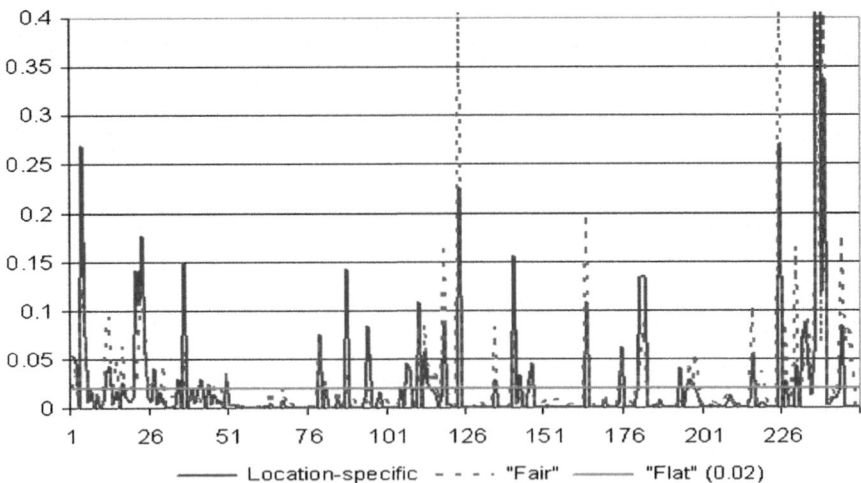

Fig. 3.7 Comparison of options: municipality-specific, "fair" and flat (0.02) premiums

For *Option 2*, Fig. 3.3 shows the distribution of municipality-specific premiums based on average damage in each municipality (or according to the municipality-specific risk). There is a prevailing number of municipalities (about 170) that have to pay 0.02–0.03 m. u., which is close to the flat rate of 0.02, as in *Option 1*. Municipalities more exposed to the risk, have to pay 0.04 and higher rates (more than 60 municipalities).

Figure 3.4 shows the distribution of the insurers' reserve (cumulated at τ within 50 years) at premiums of *Option 2*. The volume of capital is shown on the horizontal axis. The probability of insolvency (when the risk reserve accumulated up to the catastrophe is not enough to compensate incurred losses) is indicated on the right-hand ordinate axis. There is a rather high probability of "small" insolvency (values − 90, −40 occurred in about 190 and 90 simulations out of 500, as it is discussed in Sects. 3.2 and 3.7). High solvency (more than 500 m. u.) occurred in about 10% of the simulations. The size of insolvency would represent the cost to the government to cover losses uncovered by the pool. Another option may be to transfer a fraction of losses to international financial markets, as it was analyzed in Ermolieva et al. (2001).

Figure 3.5 shows the distribution of premiums for *Option 3*. According to this principle, most of the municipalities (180) have to pay close to the flat rate of 0.02–0.03 m. u. per cubic meter of a building. Rates of 0.04 and higher have to be paid by about 100 municipalities. In this case the highest premium rate is 0.5, which, in comparison to the highest rate of 1.2 of *Option 2*, is much lower. The distribution of the insurer's reserve in Fig. 3.6 indicates also the improvement of the insurer's stability: the frequency of insolvency is considerably reduced.

Figure 3.7 is very illustrative. For each municipality it shows the optional premiums to be paid: the flat premium rate of 0.02, the *Option 2* municipality-specific rate, and the "fair" premium of *Option 3*. Many municipalities in all three options have to pay the premium rate, which is about the flat rate (0.015–0.03). For quite a number of municipalities in *Options 2*, the rate significantly exceeds the flat rate. For these municipalities special attention should be given to whether they are able to pay such high premiums. *Option 3* allows us to take individual constraints on overpayments into account and work out the efficient premiums both for the insurer and municipalities.

3.7 The Solution Procedure

From the discussion in Sect. 3.4 it follows, that the welfare function $W(x, V)$ for the case study in the Tuscany region is a concave function assuming that τ and L_j^τ do not depend on x. In this case the minimization of $W(x, V)$ can be approximately solved by linear programming methods (see general discussion in Ermoliev et al. 2001). The resulting linear approximation may prove to have extremely large dimensions due to the large number of scenarios for estimating the function being optimized. The main challenge is concerned with the case when τ and L_j^τ are implicit functions of x. Then we can only use the stochastic quasigradient (SQG) methods (see Ermoliev and Wets 1988; Birge and Louveaux 1997). Let us outline only the main idea of these techniques. More details and further references can be found in Ermoliev et al. (2000, 2001).

Assume that vector x incorporates not only risk management decision variables x but also V and decisions affecting the efficiency of the sampling itself (for more detail see Pugh 1966; Ermolieva 1997). An adaptive Monte Carlo Optimization procedure (SQG method) searching for a solution minimizing $W(x)$ starts at any reasonable guess x^0. It updates the solution sequentially at steps $k = 0, 1, \ldots$, by the rule $x^{k+1} = x^k - \rho_k \xi^k$, where numbers $\rho_k > 0$ are predetermined step-sizes satisfying the condition $\sum_{k=0}^{\infty} \rho_k < \infty$, $\sum_{k=0}^{\infty} \rho_k^2 = \infty$. For example, the specification $\rho_k = 1/(k+1)$ would formally suit. Random vector ξ^k is an estimate of the gradient $W_x(x)$ or its analogs for nonsmooth function $W(x)$. This vector is easily computed from random observations of $W(x)$. For example, let W^k be a random observation of $W(x)$ at $x = x^k$ and $\bar{W}$ be a random observation of $W(x)$ at $x = x^k + \delta_k h^k$. The numbers δ_k are positive, $\delta_k \to 0$, $k \to \infty$, and h^k is an independent observation of the vector h with independent and uniformly distributed on $[-1, 1]$ components. Then ξ^k can be chosen as $\xi^k = \left[\left(\bar{W}^k - W^k \right) / \delta_k \right] h^k$. There is significant flexibility in choosing ξ^k for estimating the gradient of $W(x)$ at $x = x^k$. Some of them may lead to fast convergence, others produce slow oscillating behavior. For example, the straightforward estimation of function $\Psi(x)$ in Sect. 3.2 is time consuming. But due to formula (3.1) we can use the following procedure. Let us consider any sequence of numbers $\mu_t > 0$, $t \geq 1$, $\sum_{t=1}^{\infty} \mu_t = 1$. Step $k + 1$: choose t_k with a probability

μ_t from set $t \in \{1, 2, \ldots\}$; generate $p_k \in \lfloor \underline{p}, \overline{p} \rfloor$ and simulate claim $B_k^{t_k}$ by a catastrophe model. Calculate $\xi^k = \mu_{t_k}^{-1} \left[p(1-p)^{t_k - 1} V'_{t_k} \left(\min\{x^k, B_k^{t_k}\} - \pi t^k \right) \right] \eta^k$, where V'_t denotes the derivative of $V_t(\cdot)$, and $\eta^k = 1$, if $x^k \leq B_k^{t_k}$, and $\eta^k = 0$, otherwise. It is easy to see, e.g., from the discussion of the stochastic minimax problems in Ermoliev and Wets (1988, p. 165) that $\mu_{t_k}^{-1} \lfloor p(1-p)^{t_k - 1} V'_{t_k} \left(\min\{x^k, B_k^{t_k}\} - \pi t^k \right) \rfloor \eta^k$ is an estimate of $\Psi'(x^k)$, i.e., its expected value is $\Psi'(x^k)$. The rate of asymptotic convergence of this method (when the number of observations $k \to \infty$) is similar to other sampling based procedures.

3.8   Conclusions

Catastrophes produce losses highly correlated in space and time, which violates the law of large numbers. In this chapter we show how to address the insurability of dependent high-impact catastrophic risks by calculating conditions that would aid insurers and investors in deliberate selection of their portfolios.

We outlined the structure of a basic integrated catastrophe management model. It has a rather general form suitable for the analysis of location-specific risk reduction measures combined with different risk spreading options. It takes into account differences in vulnerability between various insurance portfolios and geographically explicit dependent losses from events occurring at different locations. It shows that the choice of decisions in the presence of catastrophic risks can be regarded as a stochastic optimization problem.

The model is illustrated with case studies, where the case of. seismic risks in the Tuscany region, Italy, is discussed in detail.

The Tuscany case is based on a comprehensive geographically distributed data set, from which we demonstrate the ability of the methodology to analyse and compare different policy options for risk mitigation and sharing.

Special attention is paid to equity and fairness for disaster loss sharing. In the case study, the model is able to analyse multiple policy options for developing insurance in an equitable and fair manner, and their effects on insurance premium and reserve funds.

References

Amendola A, Ermoliev Y, Ermolieva T, Gitits V, Koff G, Linnerooth-Bayer J (2000a) A systems approach to modeling catastrophic risks and insurability. Nat Hazard 21:381–393

Amendola A, Ermoliev Y, Ermolieva T (2000b) Earthquake risk management: a case study for an Italian region. In: Proceedings of the second Euroconference on global change and catastrophe risk management: earthquake risks in Europe IIASA, Laxenburg, Austria

Amendola A, Ermoliev Y, Ermolieva TY (2001) Earthquake risk management: a case study for an Italian region. In: Zio E, Demichela M, and Piccinini N (eds) Towards a Safer World. Proceedings of the ESREL Conference, Torino, Italy, 16–20(3):1875–1882

Arrow K (1996) The theory of risk-bearing: small and great risks. J Risk Uncertain 12:103–111

Artzner P, Delbaen F, Eber JM, Heath D (1999) Coherent measures of risk. Math Finance 9 (3):203–228

Baranov S, Digas B, Ermolieva T, Rozenberg V (2002) Earthquake risk management: scenario generator. International Institute Applied Systems Analysis, interim report IR-02-025, Laxenburg, Austria

Birge J, Louveaux F (1997) Introduction to stochastic programming. Springer series in operations research. Springer, New York

Borch K (1962) Equilibrium in a reinsurance market. Econometrica 30(3):424–444

Cummins JD, Doherty N (1996) Can insurer pay for the "Big One"? measuring capacity of an insurance market to respond to catastrophic losses. Working paper, Wharton risk management and Decision Processes Center, University of Pennsylvania, Philadelphia, PA

Dantzig GB (1979) The role of models in determining policy for transition to a more resilient technological society. IIASA distinguished lecture series http://www.iiasa.ac.at/Admin/PUBDocuments/XO-79-002.pdf

Daykin C, Pentikainen T, Pesonen M (1994) Practical risk theory for actuaries. Monographs on statistics and applied probability, vol 53. Chapman and Hall Ltd., London

Dilley M, Chen R, Deichmann U, Lerner-Lam A, Arnold M, Agwe J, Buys P, Kjekstad O, Lyon B, Yetman G (2005) Natural disaster hotspots: a global risk analysis. Disaster risk management series 5. The World Bank Hazard Management Unit, Washington, DC

Embrechts P, Klueppelberg C, Mikosch T (2000) Modeling extremal events for insurance and finance: applications of mathematics, stochastic modeling and applied probability, vol 33. Springer, Heidelberg

Ermoliev Y, Wets R (1988) Numerical techniques of stochastic optimization. Computational mathematics. Springer, Berlin

Ermoliev Y, Ermolieva T, MacDonald G, Norkin V (2000) Insurability of catastrophic risks: the stochastic optimization model. Optim J 47:251–265

Ermoliev Y, Ermolieva T, MacDonald G, Norkin V (2001) Problems on insurance of catastrophic risks. Cybern Syst Anal 37(2):220–234

Ermolieva T (1997) The design of optimal insurance decisions in the presence of catastrophic risks. International Institute Applied Systems Analysis, interim report IR-97-068, Laxenburg, Austria

Ermolieva T, Ermoliev Y, Norkin V (1997) Spatial stochastic model for optimizing capacity of insurance networks under dependent catastrophic risks: numerical experiments. International Institute Applied Systems Analysis, interim report IR-97-028, Laxenburg, Austria

Ermolieva T, Ermoliev Y, Linnerooth-Bayer J, Galambos I (2001) The role of financial instruments in integrated catastrophic flood management. In: Proceedings of the 8th annual conference of the multinational financial society, Garda, Italy

Froot K (1997) The limited financing of catastrophe risk: an overview. Harvard Business School and National Bureau of Economic Research, Cambridge

Giarini O, Louberg H (1978) The diminishing returns of technology. Pergamon Press, Oxford

Gilber C, Gouy C (1998) Flood management in France. In: Rosenthal U, Hart P't (eds) Flood response and crisis management in Western Europe: a comparative analysis. Springer, Berlin

Ginsburg V, Keyzer M (1997) The structure of applied general equilibrium models. The MIT Press, Cambridge

Grandell J (1991) Aspects of risk theory. Probability and its applications. Springer, New York/Berlin/Heidelberg

Insurance Service Office (1994) The impact of catastrophes on property insurance. Insurance Service Office, New York

Jobst N, Zenios S (2001) The tail that wags the dog: integrating credit risk in asset portfolios. J Risk Finance http://www.algorithmics.com/en/media/pdfs/thetailthatwagsthedog.pdf. Accessed Feb 2012

Konno H, Yamazaki H (1991) Mean absolute deviation portfolio optimization model and its application to Tokyo stock market. Manag Sci 37:519–531

Kunreuther H, Roth R (1998) Paying the price: the status and role of insurance against natural disasters in the United States. Joseph Henry Press, Washington, DC

Linnerooth-Bayer J, Amendola A (2000) Global change, catastrophic risk and loss spreading. GENEVA PAP Risk Insur 25(2):203–219

MacKellar L, Ermolieva T (1999) The IIASA social security project multiregional economic-demographic growth model: policy background and algebraic structure. International Institute Applied Systems Analysis, interim report IR-99-007, Laxenburg, Austria

Markowitz H (1987) Mean variance analysis in portfolio choice and capital markets. Blackwell, Oxford

Munich Re (2009) Topics geo. Natural catastrophes 2008. Analyses, assessments, positions. Munich Reinsurance Company, Munich

Munich Re (2011a) Half-year natural catastrophe review: USA. Munich Reinsurance Company, Munich

Munich Re (2011b) Topics geo. Natural catastrophes 2010: analyses, assessments, positions. Munich Reinsurance Company, Munich. http://www.munichre.com/publications/302-06735_en.pdf

National Research Council (1999) National disaster losses: a framework for assessment. Committee on Assessing the Costs of Natural Disasters. National Academy Press, Washington, DC

Petrini V (1995) Pericolosità Sismica e Prime Valutazioni di Rischio in Toscana. CNR/IRRS, Milan

Pollner J (2000) Catastrophe risk management: using alternative risk financing and insurance pooling mechanisms. Finance, private sector & infrastructure sector unit, Caribbean Country Department, Latin America and the Caribbean region. Worldbank Technical Paper, World Bank, p 495

Pugh EL (1966) A gradient technique of adaptive Monte Carlo. SIAM Rev 8(3):346–355

Rockafellar T, Uryasev S (2000) Optimization of conditional value-at-risk. J Risk 2(3):21–41

Rozzenberg V, Ermolieva T, Blizorukova M (2001) Modeling earthquakes via computer programs. International Institute Applied Systems Analysis, interim report IR-01-068, Laxenburg, Austria

Schiermeier Q (2006) Insurers' disaster files suggest climate is culprit. Nature 441:674–675

Stone JM (1973) A theory of capacity and the insurance of catastrophe risks. J Risk Insur 40:231–244, 339–355

Walker G (1997) Current developments in catastrophe modelling. In: Britton NR, Oliver J (eds) Financial risk management for natural catastrophes. Griffith University, Brisbane, pp 17–35

Chapter 4
Multiple Criteria Decision Making for Flood Risk Management

Karin Hansson, Mats Danielson, Love Ekenberg, and Joost Buurman

Abstract This paper describes a framework for multiple criteria decision making (MCDM) for flood risk management. To date, most models assessing flood impacts and coping strategies focus on economic impacts and neglect environmental and social considerations. In this paper, we develop and test an ex-ante framework for flood damage assessment, which includes a flood simulation model, a decision tool, and suggested policy strategies. Environmental and social criteria are introduced into the framework, and soft evaluations are performed in order to demonstrate the usability of the framework. The Bac Hung Hai polder in northern Vietnam serves as a case study. Results show that it is useful to add a multi-criteria perspective to flood management decisions to account for differing views and preferences. Furthermore, such a framework enables stakeholder participation in consequence analyses as well as in formulating more elaborated criteria weights.

Keywords Decision support tool • Flood risk management • Case study • Multi-criteria perspective • Simulation tool

4.1 Introduction and Background

There is increasing recognition that flood risk management should be part of a larger development plan that includes the diverse views and preferences over multiple criteria of concern to stakeholders. The overall purpose of the research presented in this paper is to develop and apply a multi-criteria framework in the

K. Hansson (✉) • M. Danielson • L. Ekenberg
Department of Computer and Systems Sciences, Stockholm University, SE-164 40 Kista, Stockholm, Sweden
e-mail: karinh@dsv.su.se

J. Buurman
Singapore-Delft Water Alliance, National University of Singapore, Singapore, Singapore

A. Amendola et al. (eds.), *Integrated Catastrophe Risk Modeling: Supporting Policy Processes*, Advances in Natural and Technological Hazards Research 32, DOI 10.1007/978-94-007-2226-2_4, © Springer Science+Business Media Dordrecht 2013

case of complex environmental and social decisions involving flood risk management. The framework includes a computer based simulation model, a set of possible policy strategies, and a decision analytical tool. The framework is particularly useful to the unique needs of policy makers in developing countries, and we focus the analysis on the Bac Hung Hai polder in northern Vietnam as a pilot study. An earlier version is discussed in (Hansson et al. 2006, 2008). A post-analysis of the initial version showed that the framework was incomplete, lacking the ability to handle multiple criteria.

4.2 The Framework

We describe a system-analytical framework for multi-criteria decision making (MCDM) for flood risk management. To date, most methods/models to determine the impact of floods and flood coping strategies have focused on economic impacts and neglected environmental and social impacts. Often, a single measure of value is used for the consequences of a decision, typically economic value. MCDM methodologies can handle several criteria at the same time, which make them particularly attractive for decisions in the public sector. Environmental and social criteria are introduced into the framework, and soft evaluations are performed in order to demonstrate the usability of the framework. For an initial version of the framework, please refer to (Hansson et al. 2008).

The framework includes a proposed approach for achieving a fair outcome involving techniques to evaluate and present the results to relevant stakeholders, including property owners, non-governmental organisations (NGOs), governments, insurers, and lending institutions/donor agencies. More than one government can be introduced into the framework, which gives the user a possibility of adding a trans-boundary perspective. The important aspects for a trans-boundary perspective are, for instance, land use changes along a river and the shared cost and maintenance of mitigation measures, both structural and non-structural. For instance, when a structural mitigation is introduced (or neglected) in an upstream country the impact is often observable in the downstream country. In order to investigate the effects of flood management strategies that take account of stakeholder views over a period of time, where flood disasters of different magnitudes may occur, a simulation approach is used in combination with a decision analytical tool.

4.3 The Simulation Model

The relative infrequency of catastrophic events and the resulting scarcity of historical loss data contribute to the difficulties of reliably estimating the risk of catastrophe losses using only standard actuarial techniques. However, by combining mathematical representations of flood occurrences with information on property values and other variables, simulation models can generate probabilistic loss estimations. In flood hazard simulation models many subsystems are closely linked, for instance, the hydrological system is dependent on the weather system and the economic system depends on the behaviour of the river system. Thus, a flood will have impact on the economy both directly and indirectly. Adding to the complexity of the model is its inherent large degree of uncertainty. The construction of a flood simulation model therefore relies on the expertise of many scientific disciplines, such as hydrology, meteorology, civil engineering, statistics, and actuarial analysis. The expertise required to construct a simulation model for flood management decisions is thus broader than the traditional actuarial domain.

The framework uses a simulation model to link the subsystems and explore their interconnections. Land use data are included in the simulation model by the use of vectors in which each cell represents a specific area of specified size. The size of the cell is specified by the geographic area under investigation. The area may contain, for example, a property or cultivated land represented by its current economic value. Each cell is identified by its location in the vector. This provides the possibility to divide land into different locations or to add a trans-boundary perspective. Moreover, if large quantities of spatial data are implemented in the model, several vectors can be introduced. Vectors containing data on soil type, elevation, inhabitants, income, cattle, etc. are also catered for in the simulation model by connecting them to the land use data. However, micro level data are often difficult to obtain and the choice of data to include in a simulation depends on the strategies under investigation.

Hydrological data on floods, structural mitigation, and damages are provided by experts. Statistical data on different types of floods of different magnitudes are implemented in the model. Based on data, a simulation model provides decision makers with the possibility to elaborate and increase or decrease variables such as damages, frequency, and strength of a flood. It should be noted that the hydrological conditions and the capacity of the river system may change over time. The flood levels for a given return period may increase due to changes in the river system, conditions of the levees, and how they are maintained. This can, for instance, cause sedimentation. Therefore, a specific function in the model allows stakeholders to alter flood data.

Monte Carlo (MC) simulations were chosen to simulate flood protection failure since they can provide reliable statistical output given a sufficiently large enough sample base, cf. (Miao et al. 2004). A MC simulation requires, for each simulation and each random variable, probability distributions. In the framework, one or several floods can occur each time period in one or more locations. Flood probability distributions, magnitudes, and locations where the floods strike are provided by hydrologists. Simulations can be set to different lengths – for instance, 20 year periods repeated 100,000 times.

Different groups of stakeholders can be represented in the simulation model. Where the focus is primarily the financing of flood risk, the stakeholders will include the government, insurers, property owners/inhabitants, NGOs, donor agencies, and lending institutions. Each stakeholder is initialised in a vector with a specific identification number. For instance, the insurance companies can be connected to a land use vector if a contract is established between a property owner and the company. One property can carry more than one insurance contract. Data on each contract is stored in a separate vector. The details on settings for contracts are established by the actual stakeholders and implemented as different policy strategies.

Each stakeholder and/or group of stakeholders is assigned different variables and wealth transformation functions described in (Hansson et al. 2008). After each time period, the wealth of individual stakeholders and stakeholder groups are calculated. That is, the financial results are saved at both micro and macro levels, for the entire geographical area, and also for each location separately (if several locations are used). Moreover, it is possible to isolate a single property owner in order to evaluate the result at the micro level.

4.4 Policy Strategies

The framework can be used in a participatory manner, for instance, as a focal point for discussions at policy meetings. For an implemented strategy to be sustainable and accepted it should preferably involve all levels of the community; cf. (Arriens 2004; Cornwall and Jewkes 1995; Hosking 2004). Discussing flood management strategies at local meetings gives the local authorities and interest groups an opportunity to participate and state their preferences.

A public policy strategy consists of multi-valued policy parameters. If we continue with the risk financing policy example, the parameters can include, for instance, the level of post-flood government compensation to victims, the premiums paid to insurers, post-disaster borrowing by flood victims, and funding for other government services such as education. As an example of a policy strategy, the model specification could include governmental compensation set to 40% of property losses and premiums to insurers set to 1%of the property value. A specific policy strategy might be advantageous to one stakeholder group but not to another. For example, a strategy that maximises an insurer's risk reserve may not be satisfactory to individual property owners. Furthermore, the government can allocate funds to non-structural measures such as education and warning systems. A threshold can be set, and if the funds exceed the threshold, damages are reduced by a percentage set by the policy maker. Moreover, if enough funds are allocated for maintaining the existing structural measures, then probabilities for damages are reduced. Thresholds are set by policy makers. Donor agencies can be introduced into the model, for instance, if the government's wealth reaches a lower boundary or if more than one flood occurs within a time period. For a more detailed description of policy variables, consult (Hansson et al. 2006, 2008).

4.5   A Decision Analytical Tool

For decision makers and stakeholders to express their preferences, and to approach a Pareto optimal flood management outcome, a decision support module is incorporated into the framework. It has been shown to be difficult for stakeholders to express their preferences unaided (Riabacke et al. 2009; Matsatsinis and Samaras 2001). The module makes it possible to analyse the decision situation and incorporate risk and uncertainty.

When assessing flood management strategies, the relative importance of the different stakeholder preferences must be considered. The result from the simulation model can be evaluated, e.g. by weighting (aggregated groups) of stakeholders. In restricted situations, it is sometimes possible to find an optimal solution for one of the parties involved. However, such a solution is problematic for several reasons. One main problem is that it would be politically impossible to make such an aggregation using a black-box approach, i.e. the stakeholders would not accept the outcome. This has been clearly shown in several interviews performed in flood settings in the Tisza basin (Vári et al. 2003; Ekenberg et al. 2003). Furthermore, even if fixed numerical weights could be introduced, there is no objective (or even inter-subjective) way of making proper final assignments. In this framework, we take account of a multitude of weights at the same time (in the form of weight intervals) and explore how they affect the outcome. Taking this approach, the results from the simulations are analysed using the decision tool and classes of weighted mean losses are calculated. This analysis incorporates sensitivity analyses of the various costs and probabilities involved.

The decision analytical module selected for the framework is based on the Delta method (Rice 1994; Danielson and Ekenberg 1998). Main features include the use of familiar concepts like weights, probabilities, and values rather than a more specialized formalism, and the possibility to use vague and imprecise input data in the form of intervals and comparisons rather than fixed numbers for input data, such as stakeholder preference importance weights, event probabilities, and values. The concepts of weights, probabilities, and values were chosen since they are well established and therefore more easily accepted by stakeholders, see, e.g., (Brouwers et al. 2002, 2004; Danielson et al. 2006). The choice is pragmatic and not to be seen as an indication of inappropriateness of other uncertainty techniques. With the simulation results as a basis, the strategies are analysed using the tool. For the evaluation of the options, aggregated data from the simulations are used. The outcomes from the simulation model are saved and the results are automatically transferred to the tool.

4.6   Adding Different Perspectives

Typically, analysts construct their analyses using a single measure of value to express the decision outcome, for example, the financial or economic value. Yet, policy makers must consider multiple perspectives that include economic, environmental, social, and political considerations. Methodologies for incorporating multiple perspectives and objectives include value trees, value functions, and trade-off analyses (Danielson et al. 2007; Larsson et al. 2005). A combination of financial and non-financial criteria can be handled by using multi-criteria decision analysis (MCDA), which makes it attractive for decisions in the public sector.

Earlier studies using the proposed framework showed that a flood management tool is helpful when used in a participatory manner, where stakeholders need guidance in choosing a flood management strategy. The framework was evaluated in a Hungarian stakeholder workshop, during a field project initiated by the Hungarian government (Danielson and Ekenberg 2012); see Chap. 14 of this volume. In this case it was important to take a multi-disciplinary approach and take account of the multi-dimensional concerns of the stakeholders, cf. (Linkov et al. 2006; ICOLD 1997; Viljoen et al. 2001). Therefore, an MCDA model is introduced into our framework and implemented in the decision analytical tool.

Critics point out, however, that MCDA evaluations are often reduced to an economic value (Phillips 2002). To avoid this problem, our framework does not require fixed numbers, but allows instead assignment of a ranking order (not fixed weights) to the criteria and a preference order over the consequences (not economic value). Moreover, adding an MCDA function does not mean that the importance of the economic aspect is lost. Since a weighting system is applied in the framework, the economic criterion can be assigned an as large weight as desired for modelling the views held within the project.

Extending the framework, two types of default criteria (other than the economic criterion), social/health and environmental, have been added to the decision evaluation model. The criteria are modelled by a criteria hierarchy and assigned weights in the form of intervals or comparisons between them. A criteria hierarchy built on these criteria is discussed below in connection with the pilot study. Note that there are now two independent sets of weights, stakeholder importance weights and perspective (criteria) weights.

4.7   Case Study: Area and Setting

In addition to the Tisza case, described in cf. (Linnerooth-Bayer et al. 2012 in Chap. 12; Brouwers and Riabacke 2012 in Chap. 13), the method was applied in a case study in Vietnam. The country is rapidly changing from a planned to a market economy. In the past, funds were allocated to a large number of water professionals and workers who maintained the flood infrastructure. With the introduction of a market economy the economic pressure on the government put this labour-intensive strategy into jeopardy.

In 2007 Vietnam's disaster risk management framework, the national strategy for natural disaster prevention, response, and mitigation to 2020, was approved by the government (CCFSC 2009). This strategy focuses on water related disasters with the objective of integrating disaster planning into socio-economic development plans. The strategy strives for combining structural and non-structural measures, giving forecasting and preparedness a central role (CCFSC 2009). The use of insurance as a risk transfer mechanism in Vietnam is close to non-existing, c.f. (GlobalAgRisk 2009; Ghesquiere and Mahul 2007). The Ministry of Labour, Invalids, and Social Affairs (MoLISA) is responsible for establishing disaster compensation policies. Moreover, a project has been initiated in order to make sure thatdonors and funding are organised and used in a coordinated way (UN 2010). The disaster management group in Vietnam is working with the objective of coordinating activities and planning in the disaster management area (VUFO-NGO 2006).

One of the largest deltas in Vietnam is the Red River Delta, which is at high risk of floods. Lives and property are threatened by annual flood events as well as extreme floods which impose a substantial burden on the communities. The Red River Delta exhibits characteristics of a region in stress: increasing numbers of floods, dense and increasing population, and a low land location. The Red River Delta, and more specifically the Bac Hung Hai polder, serves as the pilot study for our analysis.

The population in the polder is 2.8 million persons. For the pilot study, 11,200 were included, all of whom are at risk to flood. The area of the polder consists of 225,000 ha, of which most is agricultural land, and the elevation ranges from 0 to 10 m, where the highest elevation is in the northwest and the lowest in the southeast. Table 4.1 shows the value of homes in this pilot region.

Currently, in the Bac Hung Hai polder levees form the primary defence against flood disasters. In general, maintenance costs, primarily for levees, are substantial in the region. Sections are generally lying below the normal flood level and require significant efforts during flood seasons. Normally, multi-purpose upstream reservoirs function as water reservation bodies for hydroelectric plants. During the flood season their priority changes to general flood-control measures (ADRC 2005). There are also several general strategies for flood management, incorporated into development plans for the basin, such as forecasting, warning systems, and other preparedness actions (UNDP 1998, 2002).

Nine different flood scenarios were implemented. Four scenarios are described by levee failure due to seepage at four different locations in the polder (see Fig. 4.1), and four are described by overtopping, and finally, one represents no event in a particular year. Data on flood probabilities and flood damages were gathered on location, and statistics were retrieved from local authorities (see Table 4.1 for damages) (SWECO/WL 2005).

Locations and probabilities
1. Song Hong, Red River, protected by 64–80 km levees. Probability for overtopping is 4.7% and for levee breach 2.6%.
2. Song Hong 2, Red River, 80–120 km levees. Probability for overtopping is 4.2% and for levee breach 31%.
3. Suond Duong, Duong River, 0–45 km levees. Probability for overtopping is 0.5% and for levee breach 0.1%.

Table 4.1 Vulnerability variables derived from regression analysis

Type of house	Value 10^6 VND	Number of houses	Damage rate (%)
Villa	560	14	10
House with concrete frame	263	56	15
House with concrete roof	134	812	15
House with tiled roof	42	1,624	20
Thatched cottage	7	294	60

Data from (IMECH/NIAPP 2005)

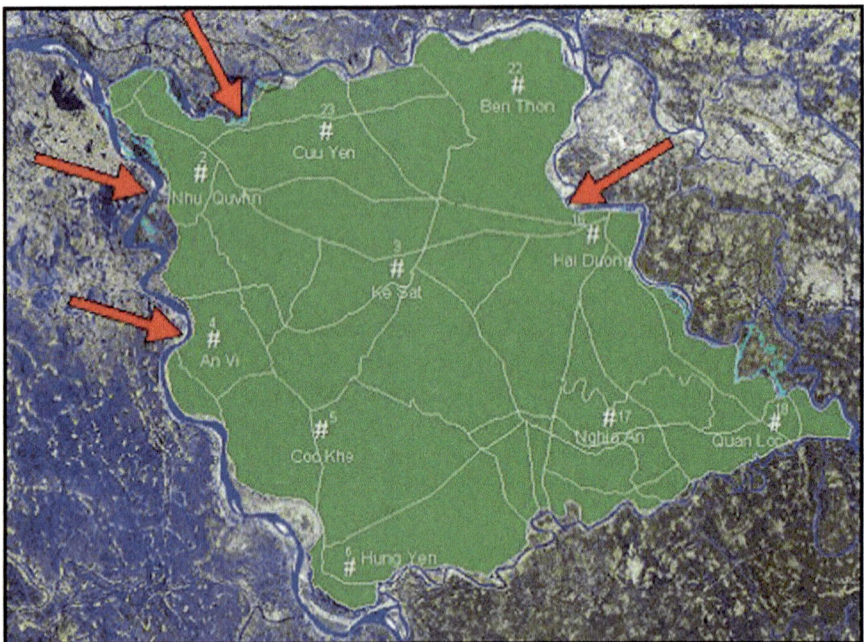

Fig. 4.1 Bac Hun Hai polder, failure locations, scale 1:200 000 (Figure from SWECO/WL 2005)

4. Song Thai Binh, at the Thai Binh River, 0–15 km levees. Probability for overtopping is 4.3%, and for levee breach 44%.

In the study, the simulations were repeated 10,000 times over 10-year periods. Floods were simulated using a Monte Carlo (MC) simulation technique, once each year per type of flood.[1] We restricted the number of floods to one flood per location a year. This gave us, for a 10 year period, more than $12 \cdot 10^{18}$ possible outcomes $((3^4)^{10})$.

---

[1] A flood event is calculated as the probability of a failure of the flood protection measures (using the probabilities for each type of event and for each location) conditional on a 100-year flood occurring. That is, the events of levee breach and overtopping may only occur if a 100-year flood event has struck at a specific location.

4.8 Criteria

The three types of default criteria in the decision evaluation model were economic, social/health, and environmental. For the non-economic criteria, the following criteria hierarchy was identified:

The environmental criterion includes three different types of consequences:

- pollution of drinking water;
- destruction of mangrove forests; and
- river pollution.

The criterion social/health aspects includes four different types of consequences:

- stress on families;
- increase in water borne diseases (including mosquito related diseases);
- historic/religious buildings destroyed; and
- snakebites.

These criteria are soft and thus more difficult to value, but they can be assigned a preference order. For each unique event and for each location, each group of stakeholders can rank criteria and assign a weight (or a weight interval) if desired. There are numerous factors that can be incorporated into a project. Below, we discuss factors included in the pilot study.

Incorporating environmental aspects are vital in a developing country. Inhabitants know the river system and its behaviour. They are often dependent on the river for daily chores such as gathering food, transportation, washing clothes, and to water crops and cattle. Therefore, a flood management strategy designed for a developed country may not be suitable for a developing country (van Ogtrop et al. 2005).

- After a flood, contaminated *drinking water* is often a problem leading to illnesses such as cholera or dysentery since the sewerage system is often underdeveloped (French and Holt 1989; MMWR 2003; Ahern et al. 2005).
- Additionally, still water after a flood (e.g., puddles, water-filled divots) can increase *diseases spread by mosquitoes*, such as malaria, dengue fever, and in some cases different forms of encephalitis (US-EPA 2006). It should be noted that mosquito borne disease transmission is related to the number of infected mosquitoes able to transmit disease and not to the total number of biting mosquitoes present in a population. Thus, the surrounding environment, such as soil type and vegetation, is related to the number of mosquitoes (Shaman and Day 2005). For instance, after a flood in 2002 in Mozambique there was an increase of malaria of 1.5–2 times (Kondo et al. 2002). It should also be noted that increased transmission of mosquito-borne diseases usually occurs several weeks after a flood (and not during a flood).
- Social protection and sickness allowance is often non-existent in developing countries. Instead, families and neighbours rely on each other. If a member of a family suffers from illness, this can be devastating to a household's economy.

During a flood family members may be separated. The Red Cross is working actively to trace missing people after disasters (Watson-Smyth 2000). To minimise the risk of such family tragedies, it is important for people living in high risk areas to be informed and educated regarding suitable escape routes and meeting points.

- During a flood, contaminated water may *pollute the river* in such a way that the environmental balance is disrupted. This affects the health and survival of poor people dependent on the river. Industries along riversides may cause serious disruption to the environment if exposed to flood. Run-off from floods can bring contaminated water into the river, poisoning the fish. Such an incident occurred in the Tisza river in 2000 (Black and Williams 2001; Linnerooth-Bayer et al. 2001; Vituki 2000), where the flood carried cyanide and heavy metal not only into the river causing fish and microbes to be poisoned, but also into wetland and flood plains along the river causing pollution to otters and birds. In more tropical regions, flood run-off can destroy coral, and the nutrients create an algae bloom producing ciguatoxin (toxic bacteria). Finally, people ingest the bacteria by eating the contaminated fish (e.g. Johnson 2005; Virola et al. 2008). Heating the fish by conventional cooking does not destroy ciguatoxin bacteria.
- *Snakebites* may cause deaths and injury after a flood. Like man, snakes seek shelter from the water in trees, houses, and on roof tops. Stressed and nervous, competing for space, the snakes bite anything close (West et. al. 2006; Huq 2004; Valcárcel 2004).
- *River mangrove forests* are vital for many reasons. They provide habitats and breeding areas for wildlife such as fish and shrimps. Mangroves protect river banks and coastal regions from erosion, and they reduce the impact of floods (Lee and Krishnapillay 2003). Thus, mangrove forests can be seen as both natural protection and as a consequence of flooding.
- *Historical buildings, ancient monuments and churches* are important to preserve (De Silva 2003; AHC 2002). They serve as symbols for the nations/areas/ villages history, religion, and heritage. Moreover, for the inhabitants in small villages, religious buildings often serve as meeting halls, and if hazard maps and escape routes are produced, this is often the place where they are kept, accessible to all villages. Furthermore, cultural and historical buildings can attract tourism and therefore be economically beneficial.

These criteria are implemented as decision trees for the evaluation procedure together with the strategies presented in Sect. 4.9. One tree is created per criteria, stakeholder, and strategy. This will be further described in Sect. 4.10 where we discuss the implementation and evaluation procedure.

4.9 Strategies

Several strategies are implemented in the simulation model. For this study, we consider only the financial aspects (reimbursement or compensation after a disaster) and the generic types of mitigation measures (structural and non-structural), which are drawn from an earlier case (Vári et al. 2003) and further elaborated to fit the needs of developing countries (Hansson and Ekenberg 2002).

- **Strategy 1**: Low government compensation in combination with structural measures.
- **Strategy 2**: The same settings as in strategy 1, but with considerably more funds to non-structural pre-mitigation measures (education and warning systems which reduce lives lost and damage).
- **Strategy 3**: The use of a catastrophe fund for compensation and maintenance, where tax money is pooled.

These have several sub-consequences under the various criteria and will be further discussed in the next section. More generally, the differences between strategies 1 and 2 are the variables and functions concerning non-structural measures (warning systems and education) in combination with higher maintenance levels of the existing levees. Higher maintenance reduces probabilities of floods occurring. Maintenance costs differ for each location based on the length of the levee, and also with reparation costs if a failure occurs. Strategies 1 and 2 use the same settings for government compensation in case of a flood, insurance rate and level of compensation from the insurer. Today, the use of insurance is close to non-existent in the Bac Hung Hai region. Therefore, insurance rates and compensation levels are set low. As in the case of Hungary (Ekenberg et al. 2003), premiums for a bundled insurance are paid from the property owners to the insurance company each year, at a certain percentage of property value where a percentage of the bundled insurance goes specifically to flood loss compensation. Therefore, in this chapter, for all strategies the poverty rate is set to 30% (GSO 2005). Households that are poor may not afford to purchase insurance. In the simulation model, it is possible to implement subsidized insurance for these households.

Strategy 1 uses less funding for warning systems and education, i.e. not enough to reach the threshold to reduce damages. In Strategy 2, however, probabilities for flood failures are reduced. Furthermore, the costs for borrowing money and/or receiving funding from donors are different for the strategies. In Strategy 2, donor assistance is provided from agencies and NGOs with different amounts depending on failure scenario, but there are no donor contributions in the case of Strategy 1. In Strategy 1, the amount of borrowed funding is assumed to be 40% higher compared to Strategy 2. If two or more failures occur within a time period, it decreases the compensation. In Strategy 2 it is assumed to decrease by 30% as compared to Strategy 1. The cost for reparations of overtopping and levee breach are the same for all strategies. In Strategy 3 we use different settings to illustrate different possible coping strategies where a mandatory fee, which goes to a catastrophe fund, is charged from all property owners. In this strategy, if an event occurs, the government compensates the property owners. No donations or aid are given and maintenance spending is enough to decrease the probabilities for failure, but no funds are provided for warning measures or education.

4.10   Analysis of Strategies

4.10.1   The Decision Tree

The decision tool automatically creates decision trees using the data provided by the simulation model. One tree is created for each group of stakeholders or an individual stakeholder if requested. In this study, each tree consists of three alternatives representing the different strategies. Each tree represents one criterion: economic, social/health, or environmental. In the economic/financial criterion tree, the alternatives lead directly to the consequences, that is, taking into consideration all possible outcomes per each time period during the simulation. The other criteria trees are created by the project members, not by the simulation tool, since they contain soft data. Here, each alternative consists of events before leading on to the possible criteria consequences.

To construct the criteria trees, the actual flood events that occurred in the simulation round are used to create event nodes (see E10 in Fig. 4.3 for an example of an event node). Thus, the occurrence of each specific flood for each year is saved in a matrix, and each unique sequence per time period is sorted out. In this pilot study, 23 unique outcomes where generated. It should be noted that the order of occurrences is taken into consideration. Thus, for instance, if two levee breaches s occur at the same location within a time period, it affects borrowing, damage rate, reparation costs, etc. The propositions of stakeholders are represented in a tree under each criterion. Each such tree contains all strategies and their consequences. This is modelled for all stakeholders and all criteria in a layered master model. This provides the decision maker with multiple layers of decision trees to evaluate all together or separately. See Fig. 4.2. Based on real probability data concerning each flood, the event node probability is calculated accordingly as

$$p(F1) \cdot (1 - p(F1))^9 \cdot p(F2) \cdot (1 - p(F2))^9 \cdot 45. \tag{4.1}$$

In Eq. (4.1), we calculate the probability for two floods occurring within a time period. $F1$ corresponds to a specific flood probability, for instance levee breach at location 1. The probability for each flood concerns one single year and not the accumulated risk. In Eq. (4.1), one type of flood F1 has occurred during the 10-year period as well as one type of flood F2. Since this concerns an entire 10-year time period, we multiply it by the binominal coefficient

$$\binom{n}{k}$$

where n corresponds to the number of possible outcomes and k corresponds to the number of floods. However, since in this case the order of the floods is important, we have taken this into consideration. Hence, there are 45 different alternative

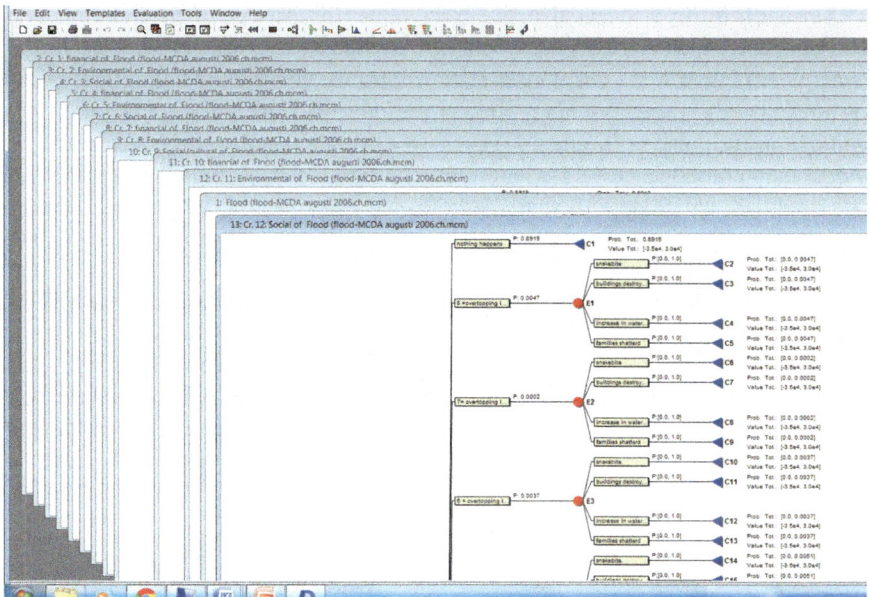

Fig. 4.2 Layers of decision trees for evaluation

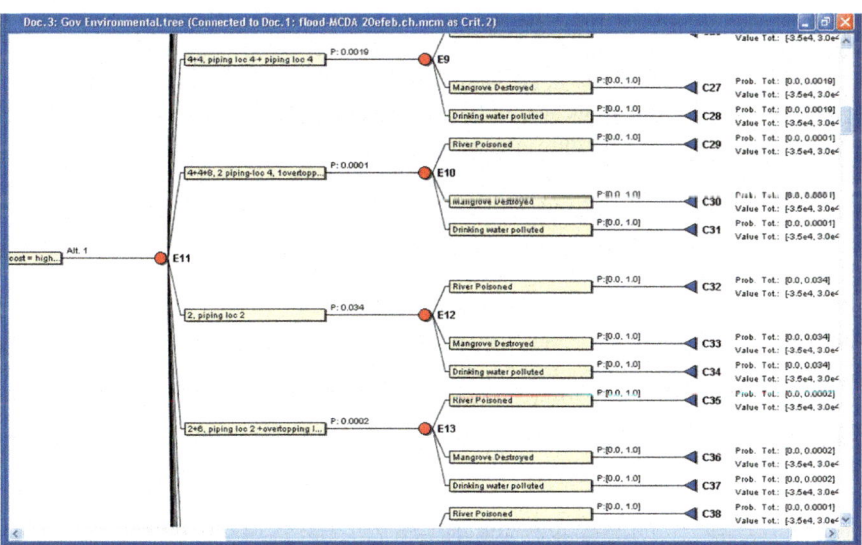

Fig. 4.3 Government environmental criteria tree

placements. The same method is used to calculate the probability of an event node
when three floods have occurred within a time period of 10 years (see Eq. 4.2).

$$p(F1) \cdot (1 - p(F1))^9 \cdot p(F2) \cdot (1 - p(F2))^9 \cdot p(F3) \cdot (1 - p(F3))^9 \cdot 120. \quad (4.2)$$

Since it is difficult to calculate the exact value (for instance of an environmental consequence), each consequence could be assigned a preference order instead. For example, in Fig. 4.3, the event node E10 denotes the consequence after three types of floods have occurred within a 10-year period, two levee breaches at location 4 and one overtopping at location 4. Here, we can state that the consequence C29 is worse than consequence C32 where only one flood, a levee breach at location 2, occurred over a period of 10 years. As a further example, we can state that for the government, it is worse if the drinking water is polluted (C30) than if the mangrove is destroyed (C31), see Figs. 4.3 and 4.4.

In Fig. 4.4, the consequences are displayed with their preference orders. This figure is an overview of all the consequences without providing details. However, it can be seen that consequence C30 is preferred to consequence C31 since C30 is located above C31. The preference order and the consequences can be different for each stakeholder. Instead of ordering the consequences in preference order, there is a possibility to assign values to the consequences, or orderings and values can be mixed as appropriate.[2]

4.10.2   Comparing Stakeholders

A conservative approach was taken for the evaluations of the social and environmental criteria. No on-site data was available. Instead, the preference order of the consequences was used as input data. No explicit numerical weights were used; instead importance rankings were used as input data. A more distinct difference between the strategies regarding economic impacts can be visualised compared to the environmental strategies (see Fig. 4.5). Throughout the analysis, the strategies are represented as alternatives in the figures; hence, Strategy 1 corresponds to Alt. 1.

In Fig. 4.5, the economic criteria for the government (left) and the property owners (right) are presented in a total ranking which presents an overview of the preference order for the alternatives using expected values. It is shown that, for the government, Alternative 3 is better than the other two alternatives and Alternative 1 is the worst. This is interesting since Alternative 2 has a larger contribution to education and warning systems than Alternative 1. Spending for Alternative 2 yields a pay-off effect in terms of less damage. Thus, a lesser amount of compensation must be provided. Alternative 3 is clearly a better choice of strategy from the government's perspective with mandatory fees contributing to a catastrophe fund. However, from the property owners' perspective, this is clearly the worst alternative financially. Furthermore, pooling money in a developing region might be difficult in reality since often there is little money to pool.

---

[2] Sometimes, these estimates are difficult to provide. Bana e Costa and Soares de Oliveira (2004) suggest a method that can be applied to such estimates, the MACBETH method (Measuring Attractiveness by a Categorical Based Evaluation Technique).

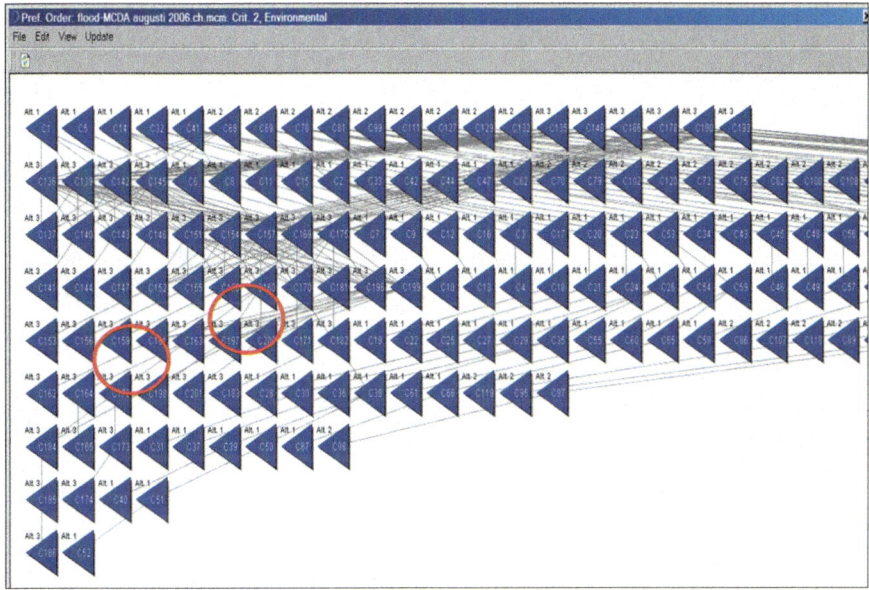

Fig. 4.4 Government environmental criteria, preference order

Fig. 4.5 Total ranking –
government and property
owners

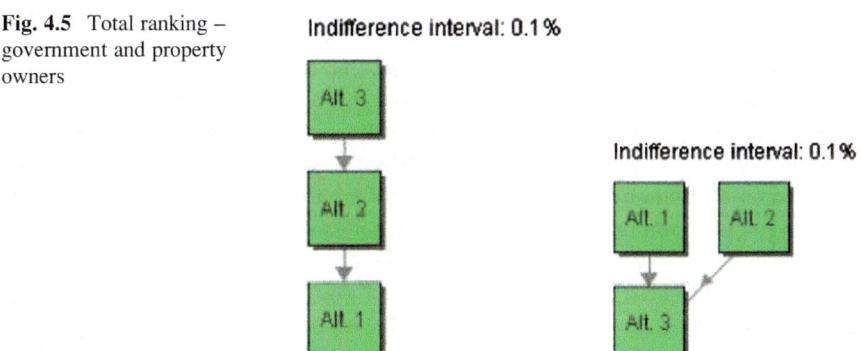

4.10.3   Weighting All Stakeholders Equally

In Fig. 4.6, the results for each criterion are shown when the preferences of all
stakeholders are weighted equally. The figure on the left illustrates the economic
criterion; the middle figure illustrates the environmental criterion, and the figure on
the right the social criterion. In the figure, all possible numerical values consistent
with the information in the decision problem are taken into consideration. The
y-axis scale is normalized to $[-1, 1]$ with -1 representing the worst and 1 the best
possible outcome. For the economic criterion, Strategy 1 is the worst strategy, and

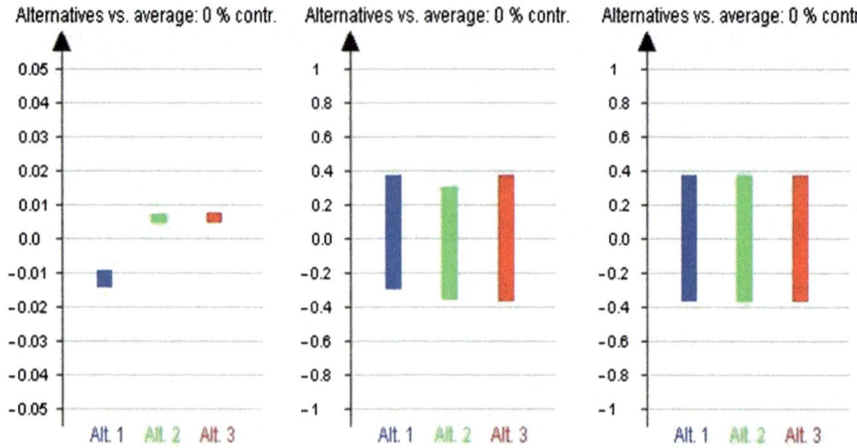

Fig. 4.6 Economic criterion (*left*), environmental criterion (*middle*), and social/health criterion (*right*)

Strategy 3 is very slightly preferable to Strategy 2. In Strategy 1, there is less investment in non-structural pre-mitigation measures, leading to increased risk and also to increased losses for property owners and the government. The largest deficit for Strategy 1 is −7.4 MUSD which was experienced by the government. In the middle, the environmental criterion is presented, and we can see that Strategy 1 is slightly preferable to Strategies 2 and 3. To the right, the criterion concerning social aspects and health is presented. The results for this criterion are similar. Thus, for this particular study, only a basic preference order is set without on-site stakeholder interviews.

Note that the settings above are for discussion purposes only and do not reflect the views of the authors. The purpose is to demonstrate the general applicability of the framework. During this study, we elaborated with several settings on the ranking of the criteria. For instance, in the settings above, we assumed that the economic criterion is the prominent rank for the government, followed by the social and the environmental criteria in no specific order. The insurer was given the same ranking, with a larger interval between the financial and the other criteria. The NGOs were given the opposite settings where the social and the environmental criteria were ranked higher than the financial. For the individuals, the financial criterion was ranked the highest followed closely by the social criterion and finally the environmental. In the modelling below, all criteria were ranked equally.

4.10.4   Pareto Optimal Solution

In Fig. 4.7, all stakeholders and all criteria are equally weighted. A Pareto optimal solution with no restrictions shows that Strategy 3 is the preferred choice. The

Fig. 4.7 All criteria and
stakeholders are equally
weighted

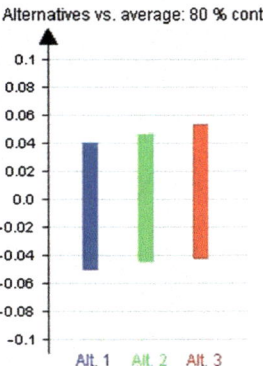

difference between Strategies 1 and 2 is small and needs further study to be conclusive. Both these are inferior to Strategy 3. However, note that Strategy 3 does not contain direct non-structural mitigation measures affecting the environmental and social criteria, but maintenance spending is enough to decrease the probabilities for failure. No further donations or aid are given. No funds are provided for warning measures or education.

Note that the described setup can also include stakeholder rankings. This is a sensitive matter and should be handled with care when modelling. Nevertheless, when the results of the possible strategies are equal, it might be meaningful to investigate the effects of an explicit stakeholder ranking.

4.11   Concluding Remarks

Improved methodologies and tools are needed to assess how government flood policies meet the multiple criteria objectives of the different stakeholders, and to evaluate different and contending policy strategies.

This paper has demonstrated a multi-criteria perspective in a framework for analysing flood management strategies. Floods are characterized not only by their economic consequences, but also by consequences that are difficult to quantify, such as loss of cultural/historical buildings or the destruction of environmentally sensitive areas. It has proven possible to include multiple criteria and combine them with the economic consequences for flood management decision making.

The study focused on a small polder in Viet Nam. It did not include fixed values for the social and environmental consequences. It was shown that assessing outcomes with preference orders for the consequences cannot provide clearly differentiated results. By assigning criteria weights, a more distinct result was obtained.

Acknowledgements This work was supported by the Swedish international development cooperation agency (Sida). Many thanks to Dr. Lars Asker for discussions regarding Matlab methods.

References

ADRC (2005) Total disaster risk management: good practices. Asian Disaster Reduction Centre, Kobe

AHC (2002) Protecting local heritage places: a guide for communities. Australian Heritage Commission, Canberra, Australia. ISBN 0642305382

Ahern M, Kovats RS, Wilkinson P, Few R, Matthies F (2005) Global health impacts of floods: epidemiologic evidence. Epidemiol Rev 27:36–46

Arriens WL (2004) Participatory processes in IWRM, first training program on IWRM and strengthening of River Basin Committees (RBC). Bangkok and Chiang Mai, Thailand, 26 July–6 August

Bana e Costa CA, Soares de Oliveira R (2004) A multicriteria model for portfolio management. Eur J Financ 10(3):198–211

Black MC, Williams PL (2001) Preliminary assessment of metal toxicity in the Middle Tisza River (Hungary) flood plain. J Soil Sediment 1(4):203–206

Brouwers L, Riabacke M (2012) Consensus by simulation: a flood model for participatory policy. In: Amendola A, Ermolieva T, Linnerooth-Bayer J, Mechler R (eds) Integrated catastrophe risk modeling: supporting policy processes. Springer, Dordrecent, Netherlands

Brouwers L, Hansson K, Ekenberg L (2002) Simulation of three competing flood management strategies – a case study. In: Ubertini L (eds) Proceedings of applied simulation and modelling (ASM). Applied Simulation and Modelling (ASM 2002), Crete, Greece, 25–28 June

Brouwers L, Danielson M, Ekenberg L, Hansson K (2004) Multi-criteria decision-making of policy strategies with public-private re-insurance systems. J Risk Decis Policy 9:23–45

CCFSC (2009) Implementation plan of the national strategy for natural disaster prevention, response and mitigation to 2020. N. 20090929, The Central Committee of Flood and Storm Control Office 6, Hanoi

Cornwall A, Jewkes R (1995) What is participatory research? Soc Sci Med 41(12):1667–1676

Danielson M, Ekenberg L (1998) A framework for analysing decisions under risk. Eur J Oper Res 104(3):474–484

Danielson M, Ekenberg L (2012) A risk-based decision analytic approach to assessing multistakeholder policy problems. In: Amendola A, Ermolieva T, Linnerooth-Bayer J, Mechler R (eds) Integrated catastrophe risk modeling: supporting policy processes. Springer, Dordrecent, Netherlands

Danielson M, Ekenberg L, Hansson K, Idefeldt, J, Larsson A, Påhlman M, Riabacke A, Sundgren, D (2006) Cross-disciplinary research in analytic decision support systems. In: Proceedings of the 28th international conference on information technology interfaces, IEEE ITI. Cavtatm, Dubrovnik

Danielson M, Ekenberg L, Larsson A (2007) Distribution of expected utility in decision trees. Int J Approx Reason 46(2):387–407

De Silva N (2003) Preparedness and response for cultural heritage disasters in developing countries. In: International symposium proceedings of cultural heritage disaster preparedness and response, Hyderabad, India, 23–27 November, pp 223–226

Ekenberg L, Brouwers L, Danielson M, Hansson K, Johansson J, Riabacke A, Vári A (2003) Flood risk management policy in the upper Tisza Basin: a system analytical approach – simulation and analysis of three flood management strategies. International Institute for Applied Systems Analysis, Interim report IR-03-003, Laxenburg, Austria

French JG, Holt KW (1989) Floods. In: Gregg MB (ed) The public health consequences of disasters. US Department of Health and Human Services, Public Health Service, CDC, Atlanta, pp 69–78

Ghesquiere F, Mahul O (2007) Sovereign natural disaster insurance for developing countries: a paradigm shift in catastrophe risk financing. Hazard Risk Management Unit, Working paper 4345, The World Bank, Washington, DC

GlobalAgRisk (2009) Designing agricultural index insurance in developing countries: a GlobalAgRisk market. Development model handbook for policy and decision makers, Lexington, KY

GSO (2005) Statistics documentation centre. General Statistics Office of Vietnam, Hanoi

Hansson K, Ekenberg E (2002) Flood mitigation strategies for the Red River Delta. In: Proceedings of the international conference on environmental engineering, Niagara Falls, Canada

Hansson K, Ekenberg L, Danielson M (2006) Implementation of a decision theoretical framework: a case study of the Red River Delta in Vietnam. In: Proceedings of the 19th international Florida AI research society conference. AAAI Press, Menlo Park

Hansson K, Danielson M, Ekenberg L (2008) A framework for evaluation of flood management strategies. J Environ Manage 86(3):465–480

Hosking A (2004) The principles of stakeholder engagement and consultation in flood and coastal erosion risk management. Halcrow Group Limited and Flood Management Division, DERFA. http://www.defra.gov.uk/environ/fcd/policy/strategy/staking.pdf. Accessed 11 June 2006

Huq S (2004) Bangladesh floods: rich nations 'Must Share the Blame', science and development network, 9 August. http://www.scidev.net/Editorials/index.cfm?fuseaction=readEditorials& itemid=125&language=1. Accessed 2 Aug 2006

ICOLD (1997) The international commission on large dams. Position paper on dams and environment. http://www.icold-cigb.net/chartean.html. Accessed 1 Aug 2006

IMECH/NIAPP (2005) Report 4. Case study of ECLAC's methodology (Draft). In: Roundtable workshop on assessing socio-economic impact of flood in Vietnam Institute of Mechanics and National Institute for Agricultural Policy and Planning, Hanoi

Johnson C (2005) Neurological channelopathy in chronic fatigue syndrome (ME/CFS). http://phoenix-cfs.org/NeurologicalChannelopathy.htm. Accessed 3 Mar 2009

Kondo H, Seo N, Yasuda T, Hasizume M, Koido Y, Ninomiya N, Yamamoto Y (2002) Post flood – infectious diseases in Mozambique. Prehosp Disaster Med 17:126–133

Larsson A, Johansson J, Ekenberg L, Danielson M (2005) Decision analysis with multiple objectives in a framework for evaluating imprecision. Int J Uncertain Fuzz Knowl-Based Syst 13(5):495–509

Lee SL, Krishnapillay B (2003) Forest genetic resources conservation and management. In: Luoma-aho T et al (eds) Proceedings of the Asia Pacific forest genetic resources programme (APFORGEN) inception workshop. IPGRI-APO, Kepong, Kuala Lumpur, Malaysia, 15–18 July

Linkov I, Satterstrom FK, Kiker G, Seager TP, Bridges TS, Gardner KH, Rogers SH, Meyer A (2006) Multicriteria decision analysis: a comprehensive decision approach for management of contaminated sediments. Risk Anal 26(1):61–78

Linnerooth-Bayer J, Vári A, Brouwers L (2012) Designing a flood management and insurance system in Hungary: a model-based stakeholder approach. In: Amendola A, Ermolieva T, Linnerooth-Bayer J, Mechler R (eds) Integrated catastrophe risk modeling: supporting policy processes. Springer, Dordrecent, Netherlands

Linnerooth-Bayer, J., Ermoliev, Y., Ermolieva, T., Galambos, I. Flood risk management in Hungary's Upper Tiszabasin: the potential use of a flood catastrophe model. American Geophysical Union, Spring meeting (2001).

Matsatsinis NF, Samaras AP (2001) MCDA and preference disaggregation in group decision support systems. Eur J Oper Res 130(2):414–429

Miao Z, Trevisan M, Capri E, Padovani L, Del Re AA (2004) Uncertainty assessment of the model RICEWQ in Northern Italy. J Environ Qual 33(6):2217–2228

MMWR (2003) Morbidity and mortality weekly report. Public health for consequences of a flood disaster 42(34):653–656. http://www.cdc.gov/mmwr/index.html. Accessed 20 May 2006

Phillips L (2002) Creating more effective research and development portfolios. London School of Economics and Political Science. http://www.catalyze.co.uk/R&D%20Portfolios.pdf. Accessed 2 Aug 2006

Riabacke M, Danielson M, Ekenberg L, Larsson A (2009) A prescriptive approach for eliciting imprecise weight statements in an MCDA process. In: Proceedings of 1st international conference on algorithmic decision theory. Venice, Italy

Rice J (1994) Mathematical statistics and data analysis, 2nd edn. Duxbury, Belmont Calif., USA

Shaman J, Day JF (2005) Achieving operational hydrologic monitoring of mosquitoborne disease. Emerg Infect Dis [serial on the internet]. http://www.cdc.gov/ncidod/EID/vol11no09/05-0340. htm. Accessed 1 Aug 2006

SWECO/WL (2005) Flood risk assessment for Bac Hung Hai Polder. 2nd RedRiver basinsector project. Report part a: water resources management. Project number 30292-03. Asian development bank (ADB),Viet Nam

UN United Nations Viet Nam (2010) Annual report. Hoan Kiem, Ha Noi, Vietnam, June

UNDP (1998) Support to the disaster management system in Vietnam. United Nations Development Programme, Hanoi

UNDP (2002) UNDP's statement on the international disaster reduction day: on disaster reduction for sustainable mountain development. J Ryan. United Nations Development Programme, Hanoi

US Environmental Protection Agency (2006) Mosquitoes and the diseases they can carry. Last updated on Tuesday, May 2nd. http://www.epa.gov/pesticides/health/mosquitoes/about_mosquitoes.htm. Accessed 20 June 2006

Valcárcel V (2004) Job opportunities arise from Colombia's floodwaters. International federation of Red Cross and Red Crescent societies. 16 November. http://www.ifrc.org/docs/news/04/04111601/. Accessed 22 July 2006

van Ogtrop F, Hoekstra F, Arjen Y, van der Meulen F (2005) Flood management in the lower Incomati river basin,Mozambique, two alternatives. J Am Water Resour Assoc 41(3):607–619

Vári A, Linnerooth-Bayer J, Ferencz Z (2003) Stakeholder views on flood risk management inHungary's upperTiszabasin. Risk Anal 23(3):585–600

Viljoen MF, du Plessis LA, Booysen HJ (2001) Extending flood damage assessment methodology to include sociological and environmental dimensions. J Water SA 27(4):517–522

Vituki Plc (2000) Summary of cyanide contamination on the Tisza River. http://www.tiszariver.com/index.php?s=results. Accessed 20 Jan 2009

Virola RA, Estrella V, Domingo EV, Amoranto GV, Lopez-Dee EP (2008) Gearing a national statistical system, towards the measurement of the Impact of climate change: the case of the Philippines. Conference on climate change and official statistics. Norway, 14–16 April

VUFO-NGO (2006) Resource centre Vietnam. Climate change working group/disaster management working group coordinator. http://www.ngocentre.org.vn. Accessed 16 Nov 2011

Watson-Smyth K (2000) Red Cross aims to reunite shattered families. Wednesday, 30 August 2000 in the independent. independent.co.uk. Accessed 2 Apr 2009

West B, Jacobs K, Breazeale L (2006) Disaster relief. Minimizing wildlife problems after a flood. Forest and wildlife research center. Mississippistate university, information sheets, IS1786

Chapter 5
Dams and Catastrophe Risk: Discounting in Long Term Planning

Tatiana Ermolieva, Yuri Ermoliev, Michael Obersteiner, Marek Makowski, and Günther Fischer

Abstract Planning dams for regional economic developments and social welfare without addressing issues related to catastrophic risks may lead to dangerous clustering of people, production facilities, and infrastructure in hazard-prone areas. The concerned region may be exposed to very large losses from the low probability-high consequence event of a dam break. Endogeneity of risks on land use decisions represents new challenges for dam development planning. In this chapter we discuss an integrated risk management model that allows the planners to deal in a consistent way with the multiple aspects, views and objectives of dam projects. We introduce the notion of robust decisions, which are considered safe, flexible, and optimal because they account for multiple criteria, risks and heterogeneities of locations and stakeholders. Specific attention is paid to the choice of proper discount factors to address long-term planning perspectives of dam construction and maintenance. We illustrate how the misperception of proper discounting in the presence of potential catastrophic events may overlook the need for dam maintenance and undermine regional safety. The proposed model can be used as a learning-by-simulation tool for designing robust regulations and policies.

T. Ermolieva (✉)
Ecosystems, Services and Management (ESM) Program, International Institute for Applied
Systems Analysis (IIASA), Schlossplatz 1, A-2361 Laxenburg, Austria
e-mail: ermol@iiasa.ac.at

Y. Ermoliev • M. Makowski
International Institute for Applied Systems Analysis (IIASA), Advanced Systems Analysis (ASA)
program, Schlossplatz 1, A-2361 Laxenburg, Austria

M. Obersteiner
International Institute for Applied Systems Analysis (IIASA), Ecosystems Services and
Management (ESM), Schlossplatz 1, A-2361 Laxenburg, Austria

G. Fischer
International Institute for Applied Systems Analysis (IIASA), Exploratory and Special Projects
(ESP), Schlossplatz 1, A-2361 Laxenburg, Austria

A. Amendola et al. (eds.), *Integrated Catastrophe Risk Modeling: Supporting Policy
Processes*, Advances in Natural and Technological Hazards Research 32,
DOI 10.1007/978-94-007-2226-2_5, © Springer Science+Business Media Dordrecht 2013

Keywords Dams and development • Equity and efficiency • Discount factors • Catastrophic risks • Multi-agent decision making • Safety constraints • Robust decisions

5.1 Introduction

The controversial debate about dams (World Commission on Dams 2000) is a debate about the very meaning, purpose and pathways for achieving development. Dams have been built for thousands of years to serve multiple purposes: to store water for hydropower, for household supply and for industrial or agricultural uses. Dams are beneficial for food security, and the expansion of infrastructure is important for micro- and macro-economic developments.

As a rule, the reliance on dams as powerful infrastructures results in the clustering of industries, people, capital, and increasing production intensity and population density in vicinal areas. On the one hand, dams may provide protection from frequent small floods; on the other hand, they create a possibility of rare but high-consequence disasters if they break. The reliance on dams in the absence of appropriate reinforcements and maintenance contributed to the Katrina hurricane disaster in New Orleans. As investigated by the American Society of Civil Engineers "A large portion of the destruction from Hurricane Katrina was caused not only by the storm itself, however, but also by the storm's exposure of engineering and engineering-related policy failures. The levees and floodwalls breached because of a combination of unfortunate choices and decisions, made over many years, at almost all levels of responsibility" (Andersen et al. 2007).

In his studies, Hirschberg et al. (1998) warned about the catastrophic risk of dams, retrieving data on frequency of dams failures and fatalities by different dam types (Fig. 5.1 and Table 5.1).

The traditional analysis of dam safety is often restricted to the use of engineering models and safety assessment approaches (Harrald 2004; RESCDAM 2001). The first guideline for dam risk assessment was introduced in 1968 after the Malpasset dam failure in France that was responsible for more than 400 injuries. After that accident, a Permanent Technical Committee for Dams (CTPB) was constituted in France, which issued a decree that made emergency plans compulsory for owners or managers of large dams, including a simulation of the "would-be-flood" scenarios, along with assessment of consequence maps showing flooded areas, wave arrival times and potential losses (Harrald 2004; RESCDAM 2001).

However, uncertainties in the assessment may cause dramatic underestimation of potential consequences and related management strategies. As admitted in RESCDAM 2001, the breach formation model, the water flow interactions with infrastructures (bridges, embankments, channels, etc.), flow in urban areas, movement of sediment and debris, are among the major uncertain processes affecting dam risk assessment. Usually, the design of dams relies on the so-called "probable maximum flood (PMF)" or "maximum limit level of risk" (Bowles 2001, 2007; IAEA 1992), which have become standard criteria over the past decades (CETS 1985; Jansen 1988). However, the PMF calculation uses a combination of facts,

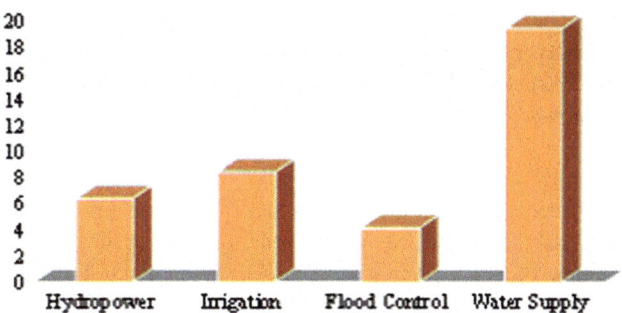

Fig. 5.1 Frequency of dams breaks, per dam-year, (scale 10^{-5}) (Hirschberg et al. 1998)

Table 5.1 Some major dam accidents and fatalities (Hirschberg et al. 1998)

Year	Name	Country	Purpose	Fatalities
1961	Panshet	India	Irrigation	1,000
1963	Vajont[a]	Italy	Hydropower	1,917
1964	Mancherla	India	Irrigation	1,000
1975	Shimantan	China	Irrig/Fl. contr.	230,000 (?)
1979	Machhu II	India	Irrig/Hydr/Syppl.	2,500
1980	Hirakud	India	Irrig/Hydr/Fl.cntrl	1,000
1993	Gouhou	China	Water supply	1,250

[a]In this case, the disaster was caused by a massive water overflow after a landslide in the lake upwards the dam remained intact

theory and expert opinions. Alternative groups of experts may arrive at different evaluation of PMF. The discrepancies in opinions arise from technical, scientific and ethical issues (CETS 1985; Jansen 1988) underlying the professional judgments, different evaluation methodologies of the estimators, and values considered in the selection of design safety objectives.[1]

As in other development programs, conflicting criteria may exist between economic efficiency and equity or ethics. Economic objectives seek to maximize benefits over costs, while equity objectives seek to find a balance between the expenditures borne by the dam owners (for dam construction and reinforcement) and the other parties, namely, those who may benefit from the dam and those who may be harmed or disadvantaged by the dam (World Commission on Dams 2000). Because of uncertainties and ambiguities, usual risk assessment methodologies are not able to determine the optimum measures to attain the economic objectives. Moreover, as we show in Sects. 5.4 and 5.5, building up public perception of dam safety on existing assessment methodologies may be dangerous. In Sect. 5.2 we limit the discussion to a few typical pitfalls. More in general, the public may have different dam acceptance criteria based on individual perception, preferences, and values (Renn 1992).

---

[1] As discussed in the following, stochastic simulation allows the analysts to consider multiple scenarios and uncertainties (see also Chap. 2 by Compton et al. in this book).

A specific emphasis of this chapter is on the management of catastrophe risks associated with possible dam breaks. Catastrophic losses have complex temporal and spatial profiles and depend on location-specific land use patterns, financial and structural mitigation decisions, and concentration of properties and population.

The design of risk management strategies calls for integrated approaches combining catastrophe models (Walker 1997) with specific decision support procedures. In Sects. 5.3, 5.4 and 5.5 we discuss an integrated modeling framework for catastrophic risk management which is being developed at the International Institute for Applied Systems Analysis (IIASA) (Amendola et al. 2000; Ermolieva et al. 2003, 2008a, b; Ermolieva and Ermoliev 2005; Compton et al. 2009) and is exemplified in other chapters of this book. In the proposed approach, the choice of decisions in the presence of catastrophic risks is supported by a spatially explicit and dynamic stochastic optimization model combining the goals and constraints of the involved agents. As safety constraints of the agents, it uses economically sound risk measures which have strong connection with the standard insolvency and stability constraints in the insurance business and Conditional Value-at-Risk (CVaR) type of risk measures (Artzner et al. 1999; Rockafellar and Uryasev 2000).

In Sects. 5.4 and 5.5, we illustrate the application of the model to a case study of catastrophic floods induced by dam breaks (see Chap. 15 by Hochrainer-Stigler in this book). Risk management decisions are being developed from the long-term perspective of welfare growth in a region when financial reserves and land use strategies for catastrophe risk management are evaluated over years. We discuss the implication of extreme events on the proper choice of discounting (Sect. 5.5). Misperceptions of discount rates may result in inadequate risk management strategies, which in turn contribute to increasing regional vulnerability and chances of catastrophes. In Sect. 5.6 we argue that the discount factors have to be linked to random arrival times of potential catastrophes ("stopping time" in our models) rather than time horizons of market interest. In general, discount rates are conditional on the degree of social commitment to mitigate risk. Random extreme events affect these rates, which alter decisions on the optimal mitigation efforts that, in turn, may change frequency and magnitude of concerned events. This endogeneity of the induced discounting restricts validity of traditional deterministic methods and calls for stochastic optimisation methods. In Sects. 5.4 and 5.5, the chapter provides insights in the nature of discounting that are critically important for developing robust long-term risk management strategies. Section 5.6 summarizes our conclusions towards directives and guidelines for integrated management of dam risks.

5.2   Ethical Goals and Constraints

Equity and ethical issues should be among the most important dimensions of dam design and maintenance. What benefits and losses the dams bring, what is a fair way of balancing them among stakeholders – these are the main questions (World Commission on Dams 2000). In many locations, the safety illusion that this powerful infrastructure creates results in intensive economic growth and concentration of properties and industries. However, without proper maintenance, the dams may deteriorate and cause major disasters. In many cases, as generally for flood protection dams, the costs for their maintenance and the responsibility for breaks are borne by central and local governments. They are the so-called social welfare maximizers responsible for the overall social wellbeing of a region. Meanwhile, the benefits and profits from dams are enjoyed by many other stakeholders. Directly and indirectly, the dams enforce and enable development not only for the current generation, but also for future generations. Therefore, mis-evaluation and misperception of the social and individual goals are at the heart of the debate around the (dis)utility of dams.

As far as risks are concerned, many existing dam assessment models are not capable of informing in an adequate manner the responsible authorities and agents (stakeholders) about the associated pros and cons. These models either focus on a very straightforward risk assessment involving primarily engineering concepts or use rather limited economic approaches, which do not reflect the nature of the dispute. Let us illustrate in the following some typical pitfalls of traditional approaches, which are overcome by the model in Sects. 5.3, 5.4 and 5.5.

5.2.1 Spatio-Temporal Heterogeneities

Temporal. The answer to ethical question about winners and losers is determined by the risk evaluation methodology. In traditional risk assessment approaches, extreme events are usually characterized by their expected recurrence periods, for example, a 1,000 year flood (e.g., associated with dam break) means a flood that occurs on average once in 1,000 years. The occurrence of a flood within a small interval t is then evaluated by a negligible probability λt, where λ reflects the event's recurrence period, e.g., $\lambda = 1/1000$. Accordingly, these events are ignored as they are evaluated as improbable during a human lifetime. This approach is frequently used in practical evaluations (ANCOLD 1998; Bowles 2007; CETS 1985; Jansen 1988). However, such an interpretation may be wrong over a long period, since the probability of a catastrophe in an interval $[0, T]$ accumulates as $1 - (1 - \lambda \Delta t)^{T/t}$ $\approx 1 - e^{-\lambda T}$. The proper assessment and management of such rare risks requires long-term perspectives. There are large uncertainties regarding the real occurrence of rare events: a 1,000-year flood may happen today, next year, or not happen at all. For example, floods across Central Europe in 2002 were classified as 1,000-, 500-, 250-, and 100-year events.

Another tendency in traditional risk assessment is to evaluate potential losses by using so-called annualization, i.e., by spreading damages, fatalities and compensation from a potential, say 50-year flood, equally over 50 years (ANCOLD 1998; CETS 1985; Jansen 1988). The main conclusion from this type of deterministic analysis is that catastrophic losses can be easily absorbed over time. However catastrophes hit as a "spike" in space and time requiring immediate financial support and adequate recovery actions.

Spatial and social. Traditional risk assessment often ignores spatial patterns of catastrophes. A general approach is to use so-called hazard maps. In most cases, these maps show catastrophe patterns that may never occur because they are developed by averaging all possible catastrophic patterns. Accordingly, social losses in affected regions are evaluated as the sum of individual losses computed on a location-by-location rather than pattern-by-pattern basis with respect to joint probability distributions. This highly underestimates the real socio-economic impacts of catastrophes dependent on simultaneous losses of assets. Furthermore losses grow exponentially with increasing network-interdependencies.

5.2.2 Multiagent Aspects

High potential impacts from a dam break call for the cooperation of various agents such as governments, insurers, investors, and individuals. The construction of a dam is a long-term project and it needs to be evaluated by taking into account the maximization of the intergenerational utility. Often, different views over the benefits of dams arise when individuals rate their instantaneous goals higher than the common-wealth, which results in dissent about the actions to be taken. Many recognize the benefits, yet, it may still be unclear how the losses associated with possible dam breaks may be shared in a fair way among the concerned agents. As estimated by many insurance companies, the losses from major dam failures cannot be borne by insurance or reinsurance companies alone. There is a need for appropriate balance between structural risk mitigation measures and risk sharing or other financial instruments involving the main concerned public and private agents. For this, the model provides a tool for a common learning from modeling and simulation.

5.2.3 Safety Constraints

For each agent, the occurrence of a disaster is often associated with his or her safety constraints, in other words, with the likelihood of some process abruptly passing individual "vital" thresholds. The design of risk management strategies therefore requires analysis and accounting of the safety constraints of the agents. For example, in the insurance industry, the vital risk process is defined by flows of premiums and claims, whereas thresholds are defined by insolvency constraints (Ermolieva and Ermoliev 2005). A similar situation arises in the control of environmental targets, in the balance of private incomes and losses, in the design of disaster management programs (Ermolieva et al. 2008a, b; see also Chap. 2 by Compton et al. in this book; Ermoliev and Hordijk 2006). Safety constraints may be represented as follows. Assume that there is a random process R_t describing the evolution of the insurer's capital reserve or accumulation of wealth in a region. A threshold is defined by a variable ρ_t. In spatial multiagent modeling, R_t and ρ_t can be large-dimensional vectors reflecting the overall situation in different locations of a region. Let us define the stopping time τ as the first moment t when R_t drops below ρ_t (e.g. because of catastrophic losses). By introducing appropriate risk management decisions, it is possible to stabilize R_t ensuring the safety constraints $P \times [R_\tau \geq \rho_\tau] \geq \gamma$, for some safety level $\gamma, t = 0, 1, 2, \ldots$ (similar to model in Sect. 5.3).

The use of safety constraints is a rather standard approach for coping with risks in insurance, finance, nuclear industries, etc. For example, typical safety regulations of nuclear plants require that the violation of safety constraints may occur only once in 10^6 years, i.e. $\gamma = 1 - 10^{-6}$ (IAEA 2001). Dams seem to have higher or comparable failure rate, see Fig. 5.1.

The ethical question about losers and winners concerns not only the evaluation of the economic benefits and costs associated with dam operation, it also relates to human and environmental values, which are often difficult to be appraised in monetary terms. The respect of safety constraints allows us to control the actions within admissible norms, say, environmental pollution, wellbeing, historical values and cultural preferences, in particular, impose regulations constraining the growth of wealth in risk prone areas. Therefore, in the model, ethical issues can be treated by evaluating the overall safety coherent with spatio- temporal goals, constraints, and indicators of involved agents – whether these are households, farmers, governments, water supplying utilities, inhabitants downstream of the dam, or insurance companies.

5.2.4   Discounting

One of the fundamental ethical parameters in the dam evaluation is the discount rate. In particular, the social discount rate reflects the level to which we discount the value of future generations' well being in relation to our own. A social discount rate of 0, for example, means we value future generations' well-being equally to our own (Ramsey 1928). Ramsey argued that applying a positive rate r to discount values across generations is unethical. Koopmans (see Weitzman 1999), contrary to Ramsey, claimed that zero discount rate would imply an unacceptably low level of current consumption.

There are several aspects of discounting to be considered in relation to dams. Traditional approaches to evaluation of dams' efficiency and safety (ANCOLD 1998; Bowles 2001; CETS 1985; Harrald 2004; ICOLD 2005; Jansen 1988; Netherlands Ministry of Housing 1989) often use principles of the so-called net present value (NPV) or modified net present value (MNPV) to justify a dam construction project. In essence, both approaches rely on the assumption that the project is associated with an expected stream of positive or negative cash flows $V_0, V_1, \ldots, V_T$, $V_t = Ev_t$ over a time horizon $T \leq \infty$. The flows may comprise several years of negative cash values reflecting the costs of construction and commissioning, followed by positive cash flow during the years without essential maintenance costs and, finally, a period of expenditures on restoration. Typically, the spatio-temporal profiles of benefits and potential dam-induced losses are not included in the evaluation. Assume that r is a constant prevailing market interest rate, then alternative dam projects are compared with respect to NPV $V = V_0 + d_1$ $V_1 + \ldots + d_T V_T$, where $d_t = d^t$, $d = (1 + r)^{-1}$, $t = 0, 1, \ldots, T$, is the discount factor, r the discount rate (Ermolieva et al. 2008a, b; ICOLD 2005; Weitzman 1999). If the NPV is positive, the project has positive expected benefits and, therefore is justifiable for implementation.

The time horizon $T \leq \infty$ and the choice of a discount rate r substantially affect the evaluation of the dam's project. Diverse assumptions about the discount rate may lead to dramatically different policy recommendations and management strategies, which may induce catastrophes and contribute to increasing vulnerability of the region.

According to Ramsey (Ramsey 1928), not facts, but ethics, are behind the choice of the discount factor and the evaluation model Lower discount rates emphasize the role of costs and benefits in the long term. The flat discount rate of 5–6% traditionally used in dam projects (Bowles 2001; CETS 1985; Jansen 1988), as Sects. 5.4 and 5.5 show, orients the analysis on a 20–30 year time horizon. Meanwhile, the explicit treatment of a 200-year disaster would require a discount rate of at least 0.5%. Section 5.5 shows that the expected duration of projects evaluated with standard discount rates obtained from traditional capital markets does not exceed a few decades and, as such, these rates cannot properly evaluate projects oriented on 1,000-, 500-, 250-, 100- year catastrophes (Ermolieva and Ermoliev 2005; Ermolieva et al. 2003; Ermoliev and Hordijk 2006).

Disadvantages of standard NPV criterion are analyzed extensively elsewhere (Chichilinskii 1997; Ermolieva et al. 2008a, b; Newel and Pizer 2000). In particular, the NPV depends on some average interest rate, which may not be implementable for evaluation of a practical project. For example, the problem that arises from the use of the expected value Er and the discount factor $(1 + Er)^{-t}$ implies additional significant reduction of future values in contrast to the real expected discount factor $E(1 + r)^{-t}$, since $E(1 + r)^{-t} >> (1 + Er)^{-t}$. In addition, the NPV does not reveal the temporal variability of cash flow streams. Two alternative streams may easily have the same NPV despite the fact that in one of them all the cash is clustered within a few time periods, but in the other one it is spread out evenly over time. This type of temporal heterogeneity is critically important for dealing with catastrophic losses which may occur suddenly in time and space.

5.2.5 Assessment vs. Robust Solutions

The assessment of risk associated with a break of a flood protection dam is usually performed in a scenario-by-scenario (what-if) manner with respect to the so-called "would-be-floods". The exact evaluation of interdependencies between all flood scenarios, potential strategies and related outcomes is impossible. It may easily run into a large number of alternative combinations. Besides, a strategy optimal against one flood may not be optimal against multiple floods. Therefore, a very important task is the design of management strategies robust with respect to all potential flood scenarios.

The underlying assumption of the robustness accounts for safety, flexibility, and optimality criteria of all agents against multiple potential scenarios of catastrophic events. Foremost, the robustness is associated with the safety constraints as described in Sects. 5.2.3 and 5.3, which deal with the Value-at-Risk considerations. The introduction of safety constraints identifies a trade-off between ex-ante (or precautionary) and the ex-post (or adaptive) measures. A balance between precautionary and adaptive decisions depends on financial capacities of the agents: how much they can invest into ex-ante risk reduction measures, such as reinforcement of dams, improving building quality or insurance coverage; and how much they are ready to spend for recovery and loss compensation if a catastrophe occurs. The future losses depend strongly on currently implemented strategies. The ex-post decisions may turn out to be much costlier, and these costs occur unexpectedly. Therefore the capacity for adaptive ex-post decisions has to be created in an *ex ante* manner.

5.3 Flood Management Model

Evaluation of measures to deal with dam breaks and induced floods is a challenging task. There is a dilemma about a proper balance between the structural and financial measures (also see Chap. 2 by Compton et al. in this book). One can argue that the increase of safety by means of investments into structural measures may avoid the need for other measures. In traditional dam management, for example, a typical goal is to reduce the probability of flooding induced by a dam break to below a certain value, the Maximum Probable Flood (MPF). Because of uncertainties in the estimates of the MPF likelihood, the investments into dam reinforcement may be essentially miscalculated.

In our case, to gain additional information on the interdependencies and ranges of potential outcomes, the analyses of risks, i.e., event probability and associated losses, are based on integrated catastrophe modeling. GIS-based Monte Carlo computer models simulate in a stochastic manner natural disasters as they may happen in reality. Catastrophe models incorporate the knowledge of the involved processes, experts and stakeholders judgments, scientific equations and variables describing them. For a flood, the latter are precipitation patterns, water discharges, river characteristics, etc. The data for developing and tuning catastrophe models are often available only on aggregated levels unsuitable for direct location specific analysis. For example, rich data on occurrences of extreme events may exist on the country level without providing information on their occurrences at specific locations. The problem of data downscaling is typical in the analysis of

precipitation and discharge curves, weather related disasters associated with changing climate, livestock epidemics modeling, missing location-specific socio-
economic data, etc. All these require the development of appropriate downscaling
procedures, which can be coupled with catastrophe modeling.

Indeed, catastrophe models aid scarce historical data with simulated samples
(scenarios) of mutually dependent catastrophic losses, which can be scaled down to
the level of individual households, municipalities, cities, or regions from various
natural hazards, e.g., floods, droughts, earthquakes, hurricanes, epidemics. These
models are becoming a key tool for land use planning, capital and industry allocation, emergency systems, lifeline analyses, and loss estimation.

Catastrophe models consist of three main modules: hazard, vulnerability, and a
multi-agent accounting system. In the case of floods, the hazard module contains
a river module, a rainfall-runoff, and a spatial inundation modules. The river
module performs calculations for a specific river. The main elements of a river
network are branches. Each branch contains several computational grid points, and
several branches are connected with nodes into a network. The mathematical model
of a branch is based on a Saint-Venant system of partial differential equations of
1-D flow mass and momentum conservation. The module transforms rainfall-runoff
water discharges into the flow dynamics using a representation of conservation
laws. Additional information on structures, reservoirs or dams along the river is
introduced. The river module may calculate the volume of discharged water into the
study region from different river branches for given heights of dams, given
scenarios of their failures or removals, and rainfall-runoff scenarios. Modeling of
breaching may be introduced as a gated weir.

The spatial GIS-based inundation module usually has a very fine resolution, say,
of 5 by 5 m grids, to capture ground elevations, soil types, water percolation
characteristics, etc. This module maps water released from the river into levels of
standing water and, thus, it estimates the area of the region affected by different
floods. For each flood event it is possible to estimate two types of maps: *Inundation*
maps that show the depth of standing water and *Duration* maps that represent how
long the water is standing on a floodplain. The module may calculate inundation
zones with inundation level of 0–2, 2–4 and more than 4 m. Duration maps show
zones covered by water, e.g., for less than 12, 12–24, 24–48 and more than 48 h.
Combination of inundation and duration maps provide time-depth-area relations,
which are used in the vulnerability module for estimation of losses caused by a
flood.

In the vulnerability module, a combination of inundation and duration maps with
so called vulnerability curves estimates potential flood losses. These can be agricultural losses depending on the inundation time, the crop and the time of the year;
property losses in buildings, depending on the depth and duration of a flood, as well
as deterioration of buildings material (wood, concrete, brick etc.). Usually, vulnerability curves are derived from historical observations. If there is no detailed GIS
information on types of buildings in a case study region, the loss estimation can be
done in relative or percentage terms. For example, in a certain sub-area of the region
the damages to wooden houses are 50% of total building value, brick houses – 40%,

and concrete houses – 10% (see Amendola et al. 2000; Ermolieva and Ermoliev 2005 and related references therein). Once the GIS distribution of building types becomes available, the relative losses can be converted into absolute ones. The vulnerability module is able to indicate changes in losses depending on changes in risk reduction measures. As an output, histograms of aggregate losses for a single location, a particular catastrophe zone, a country or worldwide can be derived from catastrophe modeling.

The integrated catastrophe model includes modules related to multiagent activities (multi-agent accounting system), e.g., farmers, infrastructure, businesses, water management, economic, financial, (re)insurance, investors, households, etc. Such multiagent accounting systems share catastrophic impacts among losses and gains of concerned agents and measure their exposure as a function of implemented strategies. These modules assist to tailor decisions accounting for complex interplay between the rainfall-runoff patterns, topography of the river, land use practices, flood defense measures in place, towards fulfillment of safety and stability constraints of agents. The outputs from a catastrophe model could show the distribution of impacts to farmers (both the distribution and across the whole sector), water authorities, urban dwellers, insurers, governmental representatives.

However, catastrophe models usually do not incorporate decision-making procedures. The integrated catastrophe management model proposed below combines catastrophe modeling and stochastic optimization procedures. Stochastic optimization (Ermoliev and Wets 1988) provides the framework necessary for incorporating interactions among decisions, agents, scenarios of catastrophes and losses into the catastrophe models (Ermoliev et al. 2000; Ermolieva and Ermoliev 2005; Ermoliev and Hordijk 2006). Adaptive Monte Carlo stochastic optimization works as follows: initial policy variables are input in the catastrophe model. The latter simulates a catastrophe and induced direct and indirect losses. The efficiency of the policies is evaluated with respect to safety (Sect. 5.2.3) performance indicators of the agents, e.g., water authorities, governments, individuals, farmers, insurers, insured, etc. If these do not fulfill the desired requirements, goals and constraints, the policies are further adjusted. In this manner it is possible to take into account complex interdependencies between patterns of catastrophes, resulting losses, policies, and safety constraints. A crucial aspect is the selection of safety constraints appropriately reflecting the risks of agents, e.g., to avoid bankruptcies.

Contrary to risk assessment, the integrated catastrophe management model estimates robust decisions, which are safe, flexible, and near to optimal, taking into account multiple criteria and heterogeneities of agents and locations affected by catastrophes. These new spatial, temporal and multi-agent distributional aspects of the integrated catastrophe management model might be the basis for policy development and implementation processes. These advantages have been explored in Amendola et al. 2000; Ermolieva et al. 2003, 2008a, b; Ermoliev et al. 2000 for the case of insurers, illustrating how the sequential optimization can improve the policies and lead them towards goals and constraints of multiple agents, e.g., on the part of insurers – to their optimal fair coverages of losses, and on the part of insured – to fair premiums, in an environment of spatial and temporal dependencies. Such improved policies suggest robust conclusions on the insurability of catastrophic risks, providing profits and stability to insurers and premium holders.

Furthermore, in contrast to models that are solely focused on simulation-based assessment of loss prevention or loss reduction measures, the multi-agent multi-objective risk-reduction and risk-sharing orientation make integrated catastrophe management models suitable for their applicability to negotiation processes. The ability of a model to clarify the results of a particular decision on the distribution of losses and benefits or to reveal potential unintended consequences allows parties to examine and identify robust policies and decisions within their own interests. The IIASA Tisza study (see Ekenberg et al. 2003; Ermolieva et al. 2003, 2008a, b, and the relevant chapters in Part III of this book) and earthquake risks management (see Amendola et al. 2000; Baranov et al. 2002; Ermolieva and Ermoliev 2005) examined the use of integrated catastrophe management models in the negotiations between stakeholders (including citizens, local and national government officials, engineers, and insurers) dealing with flood risks on the Tisza River and with policy relevant discussions of earthquake risks management for insurance legislation in Italy and Russia. The use of catastrophe models to examine the concrete impacts of different concepts of fairness as a tool in negotiations on risk may prove to be one of the most novel applications of the technique.

5.4 Case Study

We illustrate the main idea of the proposed integrated catastrophe management model by a fragment of flood risks case study on Tisza river in Hungary and Ukraine (Ekenberg et al. 2003; Ermolieva et al. 2003; Ermolieva and Ermoliev 2005) emphasizing the role of discounting for evaluation of catastrophic risks management decisions.

The main concern in the case study was the possibility of catastrophic floods due to dam breaks. The problem was to estimate the optimal reserve of a catastrophe fund to finance flood management measures including costs for dams' maintenance and loss coverage to households if dam break provokes a flood. Floods could be caused by the break of one of nine dams, which may occur as a result of a 100-, 150- and 1,000-year water discharge event into a specific river section of the region. The reliability of dams decreases without proper maintenance, which increases the chances of their failures. The system is modeled until a first catastrophic flood induced by dam break within a given time horizon. This moment is defined as the stopping time.

Let ξ be the time of a dam break. The stopping time is defined as $\tau = \xi$ for $\xi \leq T$ and $\tau = T$ for $\xi > T$. Let us denote by L_j^τ random losses at sub-location j in the study region, at time $t = \tau$, $j = \overline{1, m}$, and by π_j^t the premium rate paid by location j to the mutual catastrophe fund at time $t = 0, 1, \ldots$. Let δ_t be expenditures enabling to support the system of dikes on a specific safety level. The wealth of the fund at time τ together with a fixed partial compensation of losses $\chi \sum_j L_j^\tau$ by the government is equal to

$$W_\tau = \sum_{t=0}^\tau \left(\sum_j \pi_j^t + I_t + \chi \sum_j L_j^t - \sum_j \varphi_j^t L_j^t - \delta_t \right) \qquad (5.1)$$

where $L_j^t = 0$ for $t \neq \xi$, $0 \leq \varphi_j^t \leq 1$, is a coverage provided to location j by the catastrophe fund, $\sum_j \pi_j^t$ are premiums paid by locations to the fund, I_t is an

exogenously determined governmental investment into the fund for dike maintenance and partial coverage of losses. It is assumed that the compensation $\chi \sum_j L_j^\tau$ to

flood victims is paid by the government through the mutual fund. Indicators applied to describe the vulnerability of the flood management program are associated with insolvency of the fund, i.e., with crossing the threshold 0 by W_τ. In other words, on the probability of the event defined by inequality:

$$W_\tau < 0 \qquad (5.2)$$

The likelihood of insolvency determines the resilience of the program and, thus, the vulnerability of the region and its capability to sustain a catastrophe:

$$P[W_\tau < 0] \leq \gamma, \qquad (5.3)$$

where γ is a specified "survival" level requiring, say, that a collapse of the fund may occur only once in 10^4 years, $\gamma = 10^{-4}$.

Individuals (at locations) j receive compensation $\varphi_j^\tau L_j^\tau$ from the fund when losses occur, and pay insurance premiums π_j^t to support catastrophe mitigation program involving dams maintenance for $t = 0, 1, \ldots, \tau$.

The fairness of the flood management program is associated with the lack of overpayments by individuals determined by the indicator

$$f_j^\tau = \sum_{t=0}^\tau \left(\pi_j^\tau - \varphi_j L_j^\tau \right) > 0, \quad j = \overline{1, m}, \qquad (5.4)$$

i.e., when the level of premiums paid by a location to the fund exceeds the level of claimed losses. This is specified in the form of probabilistic constraints

$$P\left(f_j^\tau > 0 \right) \leq \rho, \quad j = \overline{1, m}, \qquad (5.5)$$

where ρ ensures the fairness by allowing overpayments with a reasonable likelihood, say, only once in 100 years, $\rho = 0.01$. Sustainable performance of the fund depends on the inflow of premiums determined by the willingness of individuals to accept the premiums, which, in turn depends on the probability of premiums overpayments (5.5).

Each agent in the model is concerned with maximizing his wealth and minimizing risks of insolvency which to major extent determine the feasibility and the demand for insurance. The main goal of the program can be formulated as the minimization of uncovered losses to households together with governmental aid and investments subject to (5.3) and (5.5):

$$F(x) = E\left[\sum_j (1 - \varphi_j^\tau) L_j^\tau + \chi \sum_j L_j^\tau + \sum_{t=0}^\tau I_t\right]. \tag{5.6}$$

It can be shown that minimization of function

$$F(x) + E\mu_0 \max\{0, W_\tau\} + E \sum_j \mu_j \max\{0, f_j^\tau\} \tag{5.7}$$

5.4.1 Discounting and Robust Decisions

The traditional risk assessment analysis often relies on discounting future losses and gains to their present values. These evaluations are used to justify risk management decisions for examples such as construction and maintenance of flood protection dams. The misperception of proper discounting rates critically affects evaluations and may be rather misleading. A common approach is to discount future costs and benefits using a geometric (exponential) discount factors as $V = \sum_{t=0}^\infty d_t V_t$, where $d_t = (1 + r)^{-t}$, r is a discount rate, and $V_t = Ev_t$ is an expected cash flow for some random variables v_t, $t = 0, 1, \ldots$. According to this standard approach, the minimization of function (5.6) has to be replaced by minimization of expected present value

$$V(x) = \sum_{t=0}^\infty d_t V_t, \tag{5.8}$$

where

$$V_t = E\left[\sum_j (1 - \varphi_j^t) L_j^t + \chi \sum_j L_j^t + I_t\right]. \tag{5.9}$$

The infinite time horizon in $V(x)$ creates an illusion of truly long-term analysis. The choice of discount rate r as a market interest rate within a time horizon of existing financial markets is well established. The following simple fact shows that the standard discount factors obtained from markets orient policy analysis only to few decades, which is not appropriate for catastrophic impacts.

Let $p = 1 - d$, $d = (1 + r)^{-1}$, $q = 1 - p$, and let η be a random variable with the geometric probability distribution $P[\eta = t] = pq^t$, $V_t = Ev_t$, where random variables v_t are independent of $v_{t+1}, v_{t+2}, \ldots$. It is easy to see that

$$\sum_{t=0}^\infty d_t V_t = E \sum_{t=0}^\eta v_t, \tag{5.10}$$

where $d = d^t, t = 0, 1, \ldots$. This is also true for general discounting $d_t = (1 + r_t)^{-t}$ with time varying discount rate r_t, where the stopping time is defined as $P[\eta \geq t] = d_t$. From (5.10) follows, that (5.8) can be written as undiscounted random sum with a new stopping time s instead of τ:

$$V(x) = E\left(\sum_j (1 - \varphi_j^s)L_j^s \mid \chi \sum_j L_j^s\right) + \sum_{t=0}^s I_t, s - \min(\eta, \tau), \qquad (5.11)$$

i.e., standard criterion (5.10) unlike proposed undiscounted criterion (5.11) orients the long-term evaluation of risk management decisions on time horizons not exceeding random horizon η associated with market interest rate r.

The expected duration of η, $E\eta = 1/p = 1 + 1/r \approx 1/r$ for small r. The same holds for the standard deviation $\sigma = \sqrt{q}/p$. Therefore, for the interest rate of 3.5%, $r \approx 0.035$, the expected duration is $E\eta \approx 30$ years, i.e., this rate orients the policy analysis on an expected 30-year time horizon. Certainly, this horizon has no relation to how society has to deal with, say, an expected 100-, 150-, 300-, 1,000- year catastrophe flood. It is essential that the proposed undiscounted criterion (5.6) links the long term evaluation of risk management decisions to horizons τ of potential catastrophic events rather than horizons of market interests. In this sense, the use of stopping time and undiscounted criterion (5.11) instead of standard discounted criterion (5.10) leads to robust flood management decisions, which are sensitive to rare catastrophic events under the analysis. That is, in the presence of catastrophic events, robust decisions are fundamentally different from decisions ignoring them.

5.5   Risk Communication, Public Perception and Participation

Communication of dam risks with the public plays a crucial role in ensuring that a community assumes the correct attitude towards dams, understands the risks posed by a dam versus benefits that it offers, and promotes efforts for better dam maintenance and regulations for land use planning and control (Amendola 2001).

A correct risk communication may build upon a set of robust strategies derived with an integrated management framework described above. The model does not provide exact remedies and answers as to the costs and benefits of each individual alternative, rather it identifies the preference structure for the actions. Let us illustrate how very simple order of actions having relation to discounting may contribute to increasing the safety of a region over a long time period.

5.5.1   Intertemporal Inconsistency of Discounting

In the case of a flood management program outlined in Sect. 5.3 (Project Proposal 2000), the problem of dam maintenance is considered to be a community responsibility. Public awareness is characterized by the perception of actions required to maintain a system of dams. In the simplest case, this may be reflected in the choice of an appropriate discounting factor. There may be several major differences in discounting approaches. For example, underestimation of dam risks may lead to the choice of the so-called time consistent discounting.

Time consistent discounting means that the evaluation of a project today ($t = 0$), will have the same discount factor as the evaluation of the same project after any time interval $[0, T]$ in the future. Hence, independently of waiting time t delayed until the implementation of the actions, the probability of the stopping time occurrence (dam break) at $t + s$ is the same, as at the initial time moment $t = 0$.

For example, traditional geometric or exponential discounting used in risk assessment, $d_t = d^t = e^{(\ln d)t} = e^{-\lambda t}$, $\lambda = -\ln d$, defines time consistent preference:

$$\sum_{t=0}^{\infty} d^t V_t = V_0 + dV_1 + \ldots + d^{T-1} V_{T-1} + d^T [V_T + dV_{T+1} + \ldots].$$

This is also connected with the geometric probability distribution of the discount related stopping time τ in (5.2), (5.3):

if $P[\tau \geq t] = d^t$, $0 < d < 1$, then $P[\tau = t] = d^t - d^{t+1} = (1 - d)d^t$, $t = 0, 1, \ldots$.

In reality, dams wear out and the probability of their failure changes with time, i.e., the discount factors have explicit time-dependent structure. The time inconsistency of delayed projects requires appropriate adjustments of discount factors for projects undertaken later rather than earlier. If a community is responsible for dam maintenance, as in the program outlined in Sect. 5.3, the misperception of this inconsistency may provoke increasing vulnerability and catastrophic losses. Let us consider typical scenarios of such developments.

5.5.2 Commitment to Actions

The analysis of social commitments to mitigate risks would require sociological studies which are outside the scope of our work. We only exemplify possible courses of (in-)actions after the model by Winkler (2006) who has defined a naïve, a sophisticated and a committed (ideal) community. The main differences between these communities and how their inappropriate choice of discounting can result in wrong decisions are studied in Ermolieva et al. 2008a, b. Below we summarize the main idea. Let us assume that planning of actions to mitigate flood risk has a fixed 100-year horizon T, in which three communities, the naïve, the sophisticated, and the committed, live and plan for coping with the catastrophic losses that may occur due to break of a dike from 150-year flood with a time consistent geometric probability distribution. The communities are able to maintain the reliability of dams by collecting money in a catastrophe fund for further investing them into dam retrofitting. But, depending on their perception of risk profiles or induced discounting, the results may be dramatically different.

The current generation of the *Naïve Community* is aware of a possible catastrophe but it has a misleading view on the catastrophe, namely, if the catastrophe has not occurred in the previous generation the community believes that there is the same probability that it will not occur within the current generation. Thus, it relies on the geometric probability distribution and fails to take into account the increasing probability of a dike break due to aging processes. The risk profiles, time preferences, premiums, and retrofitting actions are not adjusted towards the real escalating risks. In a similar way behave the other generations of the Naïve community. The plans are never implemented and the view on a catastrophe remains time-invariant despite dramatic increase of risk.

The *Sophisticated Community,* even if it has a correct understanding of the time-inconsistent discounting induced by the deteriorating dams, postpones the decisions because they value much more the present welfare and prefer to pay larger premiums delaying the actions. Due to these delays, the risk burden is increasingly shifted to the next generation, calculated premiums become higher and higher. If a catastrophe occurs, the region will also not be prepared to cope - with losses as ex-ante risk financing measures are not implemented. The unpreparedness of these communities can be explained by their misperception of risks, and, the lack of commitment to act.

In contrast, the *Committed Community* is able to implement decisions, they understand that the delays in actions may dramatically affect individuals and the growth of societies as a whole. Individuals could be better off if their consumption options were limited and their choices constrained by anticipated risks. As a direct consequence of the committed actions, the premiums that the members of community pay for coping with catastrophes become much lower than those of the sophisticated one.

5.6 Some Conclusions for Policy Evaluations

Explicit full representation of dam break risks represents new challenges for dam development planning. Pure engineering approaches, risk assessment strategies and tolerability curves are not sufficient to reflect complex interdependencies between technical and societal (ethical) criteria. According to a risk tolerability approach, the cost-effectiveness of risk reduction measures relies on the ratio of the annualized costs of risk-reduction measures divided by the annualized losses. This approach is typical to currently existing cost-benefit analysis. However, the major challenge for dam evaluation is the ability to account for the endogeneity of risks affecting large territories and to design robust strategies to simultaneously improve the well-being of multiple agents across generations. This requires the development of new type of spatio-temporal integrated models and decision support procedures, where evaluation of dam safety is connected with the evaluation of the overall safety of socio-economic and developments. The safety constraints are represented in the form of vital thresholds. In regulations of insurance business this type of constraints is called Value-at-Risk measure indicating that an insurer may become bankrupt only within the time interval specified by the board of insurers. In stochastic optimization, these are called probabilistic or chance constraints. They introduce implicit risk aversion in the selection of risk management strategies and, therefore, stress the importance of ex-ante mitigation measures.

Rare catastrophes set a restriction on the choice of discounting for evaluation of dam projects. Traditional discounting rates, based on a lifespan of current financial markets, set evaluation horizons only for 20–30 years, which may dramatically underestimate potential dam break losses and contribute to increasing vulnerability of the society. For the evaluation of "catastrophic" projects, say, long-term investments into a dam system, the discount factors have to be relevant for the expected horizons of potential catastrophes.

Our conclusion is that the integration of multiple models, concepts and views within a catastrophe management model is feasible and yields valuable insights into the robustness of the different mitigation alternatives. Major challenges, though, are due to the different perception and representation of the dam risks by agents and different disciplines. The studies suggest that integration of models and views is not simply a sum of individual components. The development of a truly integrated model should be at the basis of a societal process of model-based learning-by-simulations and communication of results. The model becomes a truly valuable tool if academic experience and expertise provides rigorous proofs and examples of where and how it can be used and what are the related consequences.

Acknowledgements The authors are grateful to Aniello Amendola for valuable suggestions and references that he provided for improving the chapter.

The chapter is partially based on a paper published in the book "Dam-break Problems, Solutions and Case Studies", Wrachien and Mambretti (eds), WIT Press, Southampton, UK, pp. 241–272. Permission by the Publisher to use previous material is gratefully acknowledged.

References

Amendola A (2001) Integrated management of technological disasters, first annual IIASA-DPRI meeting on integrated disaster risk management: reducing socio-economic vulnerability. IIASA, Laxenburg, Austria, 1–4 Aug 2000. http://www.iiasa.ac.at/Research/RMS/dpri2001

Amendola A, Ermoliev Y, Ermolieva T (2000) Earthquake risk management: a case study for an Italian region. In: Proceedings of the second Euroconference on global change and catastrophe risk management: earthquake risks in Europe. IIASA, Laxenburg, Austria 6–9 July 2000

ANCOLD (1998) Guidelines on risk assessment. Working group on risk assessment. Australian National Committee on Large Dams, Sydney, NSW, Australia

Andersen CF, Battjes JA, Daniel DE, Edge B, Espey W, Gilbert RB, Jackson TL, Kennedy D, Mileti DS, Mitchell JK, Nicholson P, Pugh CA, Tamaro G, Traver R, Buhrman J, Dinges CV, Durrant JE, Howell J, Roth LH (2007) The New Orleans hurricane protection system: what went wrong and why. American Society of Civil Engineers, Reston

Artzner P, Delbaen F, Eber JM, Heath D (1999) Coherent measures of risk. Math Financ 9 (3):203–228

IAEA (International Atomic Energy Agency) (1992) The role of probabilistic safety assessment and probabilistic safety criteria in nuclear power plant safety. Safety series no. 106. IAEA, Vienna, Austria

IAEA (International Atomic Energy Agency) (2001) Applications of probabilistic safety assessment (Psa) for nuclear power plants IAEA-TECDOC-1200, Vienna, Austria

Baranov S, Digas B, Ermolieva T, Rozenberg V (2002) Earthquake risk management: a scenario generator. International Institute Applied Systems Analysis, Interim report IR-02-025, Laxenburg, Austria

Bowles D (2001) Evaluation and use of risk estimates in dam safety decisionmaking. In: Proceedings of the united engineering foundation conference on risk-based decision-making in water resources IX, 20-year retrospective and prospective of risk-based decision-making, Santa Barbara, California. American Society of Civil Engineers, Aug 2001

Bowles D (2007) Tolerable risk of dams: how safe is safe enough? In: Proceedings of dams annual conference, Philadelphia, PA, Mar 2007

Chichilinskii G (1997) What is sustainable development? Land Econ 73:467–491

Commissions on Engineering and Technical Systems (CETS) (1985) Safety of dams: flood and earthquake criteria. http://www.nap.edu/openbook.php?record_id=288&page=11

Compton KL, Faber R, Ermolieva TY, Linnerooth-Bayer J, Nachtnebel H-P (2009) Uncertainty and disaster risk management: modeling the flash flood risk to Vienna and its subway system. International Institute Applied Systems Analysis, IIASA research report RR-09-002, Laxenburg, Austria

Ekenberg L, Brouwers L, Danielson M, Hansson K, Johannson J, Riabacke A, Vári A (2003) Flood risk management policy in the upper Tisza Basin: a system analytical approach. International Institute Applied Systems Analysis, Interim report IR-03-003, Laxenburg, Austria

Ermoliev Y, Hordijk L (2006) Global changes: facets of robust decisions. In: Marti K, Ermoliev Y, Makowski M, Pflug G (eds) Coping with uncertainty: modeling and policy issue. Springer, Berlin/New York

Ermoliev Y, Wets R (1988) Numerical techniques of stochastic optimization. Computational mathematics. Springer, Berlin

Ermoliev Y, Ermolieva T, MacDonald G, Norkin V (2000) Stochastic optimization of insurance portfolios for managing exposure to catastrophic risks. Ann Oper Res 99:207–225

Ermolieva T, Ermoliev Y (2005) Catastrophic risk management: flood and seismic risks case studies. In: Wallace SW, Ziemba WT (eds) Applications of stochastic programming. MPS-SIAM Series on Optimization, Philadelphia

Ermolieva T, Ermoliev Y, Fischer G, Galambos I (2003) The role of financial instruments in integrated catastrophic flood management. Multinatl Financ J 7(3&4):207–230

Ermolieva T, Ermoliev Y, Fischer G, Makowski M (2008a) Induced discounting and risk management. International Institute Applied Systems Analysis, Interim report IR-07-040, Laxenburg, Austria

Ermolieva T, Ermoliev Y, Fischer G, Makowski M (2008b) Integrated modeling for management of catastrophic risks: spatial stochastic optimization model. In: Knopov PS, Pardalos PM (eds) Simulation and optimization methods in risk and reliability theory. Nova, New York

Harrald J (2004) Review of risk based prioritization/decision making methodologies for dams. US army corps for engineers, The George Washington University, Institute for Crisis, Disaster, and Risk Management, Washington, DC, 29 Apr

Hirschberg S, Spiekerman G, Dones R (1998) Severe accidents in the energy sector. PSI Report Nr. 98–16, Villigen, Switzerland

ICOLD (2005) Risk assessment in dam safety management: a reconnaissance of benefits, methods and current applications. International Commission on Large Dams (ICOLD) Bulletin 130, Paris

Jansen R (1988) Advanced dam engineering for design, construction, and rehabilitation. Springer, New York. ISBN 0442243979, 9780442243975

Netherlands Ministry of Housing (1989) Physical planning and environment. Dutch national environmental policy plan – premises for risk management, second chamber of the states general, session 19881989, 21 137, No. 5

Newel R, Pizer W (2000) Discounting the distant future: how much do uncertain rates increase valuations? Economics technical series, Pew Center on Global Climate Change, Arlington, VA, USA. htpp://www.pewclimate.org

Project Proposal (2000) Flood risk management policy in the upper Tisza Basin: a system analytical approach. International Institute for Applied Systems Analysis, Laxenburg, Austria

Ramsey F (1928) A mathematical theory of savings. Econ J 138:543–559

Renn O (1992) Concepts of risk: a classification. In: Krimsky S, Golding D (eds) Social theories of risk. Praeger Publishers, Westport, pp 53–82

RESCDAM (2001) Development of rescue actions based on dam-break flood analysis. Community action programme in the field of civil protection, Finnish Environment Institute, final report June 1999–March 2001

Rockafellar T, Uryasev S (2000) Optimization of conditional value-at-risk. J Risk 2:21–41

Walker G (1997) Current developments in catastrophe modelling. In: Britton NR, Oliver J (eds) Financial risks management for natural catastrophes. Griffith University, Brisbane, pp 17–35

Weitzman M (1999) Just keep on discounting. But . . . . In: Portney P, Weyant J (eds) Discounting and intergenerational equity. Resources for the future (RFF) Press, Washington, DC

Winkler R (2006) Now or never: environmental protection under hyperbolic discounting. Working paper 06/60, CER-ETH Center of Economic Research at ETH, Zurich

World Commission on Dams (WCD) (2000) Dams and development: a new framework for decision-making. The report of world commission on dams. Earthscan Publications Ltd., London/Sterling

Part II
Disasters and Growth: Modeling and Managing Country-Wide Catastrophe Risk

Chapter 6
Modeling Aggregate Economic Risk: An Introduction

Reinhard Mechler

Abstract Not only can disasters directly cause immense human suffering and loss, they may also lead to large medium- and long-term indirect microeconomic (household, business level) as well as macroeconomic (nation-wide) consequences. In this second section of the book, we focus on the aggregated or macroeconomic impacts of disasters, which include effects on gross domestic product, consumption, savings, investment and inflation, as well as the reallocation of resources to relief and reconstruction. Based on statistical and model-based analyses, studies have shown that these impacts can be significant in many instances.

Keywords Country level disaster risk • Macroeconomic analysis • Economic growth • Fiscal vulnerability • Multi-risk assessment

6.1 Disasters and Economic Development: The Empirical Evidence

Not only can disasters directly cause immense human suffering and loss, they may also lead to large medium- and long-term indirect microeconomic (household, business level) as well as macroeconomic (nation-wide) consequences. In this second section of the book, we focus on the aggregated or macroeconomic impacts of disasters, which include effects on gross domestic product, consumption, savings, investment and inflation, as well as the reallocation of resources to relief and reconstruction. Based on statistical and model-based analyses, studies have shown that these impacts can be significant in many instances.

R. Mechler (✉)
Risk, Policy and Vulnerability (RPV) Program, International Institute for Applied Systems Analysis (IIASA), Schlossplatz 1, A-2361 Laxenburg, Austria
e-mail: mechler@iiasa.ac.at

A. Amendola et al. (eds.), *Integrated Catastrophe Risk Modeling: Supporting Policy Processes*, Advances in Natural and Technological Hazards Research 32,
DOI 10.1007/978-94-007-2226-2_6, © Springer Science+Business Media Dordrecht 2013

As an introduction we discuss different modeling techniques for estimating indirect macroeconomic disaster losses and their advantages and challenges. One main challenge is the difficulty of accounting for disaster risk using probabilistic approaches. This challenge is addressed and forms the backbone of the following three chapters of this section.

The impact of disasters on aggregate economic performance and development has been examined by several studies over the last four decades based on empirical and statistical analysis as well as modeling exercises (for an overview, see Okuyama 2009). While early studies focused primarily on developed countries and the sectoral and distributional impacts of disasters, recent studies have placed more emphasis on macroeconomic impacts in developing countries. For developed countries, the studies generally find very limited aggregate macroeconomic impacts, but important regional economic and distributional effects (Okuyama 2009). In contrast, in developing countries disasters have been found to lead to significant adverse macroeconomic outcomes that affect the pace and nature of economic development (see Otero and Marti 1995; Benson and Clay 2004; ECLAC 2003; Charveriat 2000; Mechler 2004; Kellenberg and Mobarak 2008; Hochrainer 2009; Raddatz 2007; Noy 2009; Cavallo and Noy 2009). However, as a recent review by Handmer et al. (2012) suggests, there is only *medium confidence* in the findings overall, especially since a very few studies have found positive effects (Albala-Bertrand 1993; Skidmore and Toya 2002). This lack of confidence can be attributed to several factors: the problem of identifying a systematic and robust GDP counter-factual in terms of GDP projections without disaster events (against which actual GDP including disaster effects can be compared); difficulties in accounting for informal sector effects and financial inflows (insurance and aid); and, importantly, the problem that national accounting generally measures flows rather than stocks. The latter point means that reconstruction shows up positively in national statistics, whereas the destruction does not enter the national accounts.

Keeping these reservations in mind, there is consensus that macro effects are more pronounced in lower income countries (Mechler 2004; Lal et al. 2012). Handmer et al. (2012) suggest that developing countries exhibit higher economic vulnerability due to their dependence on natural capital and disaster-sensitive activities (such as tourism), lack of developed risk assessment processes, and knowledge and implementation gaps regarding techniques for responding to disasters, including preparedness, financing, information, risk management and governance. Countries with one or more of the following characteristics may be particularly at risk of significant macroeconomic consequences (Mechler 2004): (i) high natural hazard exposure; (ii) economic activity clustered in a limited number of areas with key public infrastructure exposed to natural hazards; and (iii) tight constraints on tax revenue and domestic savings, shallow financial markets, and high indebtedness with little access to external finance.

6.2 Modeling Macroeconomic Impacts of Disasters

Findings on indirect disaster losses are in part based on modeling exercises, for which there is a substantial, yet very heterogeneous body of research. Analysts investigating indirect losses have made use of input-output models, social accounting matrices, computable general equilibrium models and economic growth frameworks. Rose (2004) reviews some of the important aspects of economic modeling, and Okuyama (2009) discusses details regarding their pros and cons. The discussion by Okuyama is taken as a point of reference and extended in Table 6.1.

The input-output (I-O) model is the most commonly employed modeling framework for estimating economic effects of disasters (Okuyama 2009). I-O models document the interdependencies within an economy and the ensuing disruption due to disasters. Their advantages are that they are directly based on observable data (usually national accounting statistics) as well as their simplicity. The major downsides include their rigid structure with respect to input and import substitutions, a lack of explicit resource constraints, and a lack of responses to price changes. As I-O models are rather rigid and focus on the short term, they may underestimate disaster impacts (see Okuyama 2009). I-O modeling has been used in conjunction with transportation network models (Gordon et al. 1998; Cho et al. 2001; Sohn et al. 2004), lifeline network models (Rose 1981; Rose et al. 1997; Rose and Benavides 1998), and comprehensive disaster assessment models, such as HAZUS (Cochrane et al. 1997).

Social accounting matrices (SAM) are a variant of I-O models and estimate the macroeconomic follow-on effects of disasters on production, savings and consumption activities of different agents. The SAM approach is well suited for handling the distributional impacts of a disaster, for instance, for evaluating the equity considerations of public policies that aim at reducing losses. Like I-O models, however, the SAM technique is rigid in terms of its coefficient structure and thus tends to provide upper bounds for the estimates (given that the coefficients are assumed fixed over time). SAMs have been used to examine the distribution of impacts across private and public sectors connected to lifeline analysis (effects on distribution networks such as for electricity and water) and to examine the effects across regions (Cole 1995, 1998, 2004; Ellson et al. 1984).

Computable general equilibrium (CGE) models show how disasters impact regional or developed economies through sectoral and price changes. They can represent non-linear effects as well as responses to price changes. As these models have (originally) been intended for long-run equilibrium analysis and assume adaptability of economic agents, they may lead to underestimating economic impacts. CGE models have been used for assessing the potential effects of catastrophic earthquakes on regional economies, and have also been linked to lifeline analysis (Boisvert 1992; Brookshire and McKee 1992; Rose and Guha 2004; Rose and Liao 2005).

Table 6.1 Modeling techniques used to assess the macroeconomic effects of disaster risk

Modeling technique	Questions addressed	Pros	Cons	Applications
Input-output (I-O)	Interdependencies within an economy	Based on actual data, simplicity	Rigid structure with respect to input and import substitutions, lack of explicit resource constraints, and lack of responses to price changes	In conjunction with transportation network models, lifeline network models, and comprehensive disaster assessment models
Social Accounting Matrix (SAM)	Higher-order effects across different socio-economic agents and activities	Distributional impacts	Rigid coefficients, provides upper estimates bounds	Distribution of impacts across private and public sectors
Computable General Equilibrium (CGE)	Sectoral and price effects within a regional or developed economy	Non-linear effects, response to price changes, can incorporate input and import substitutions, and can explicitly handle supply constraints	Rather intended for long-run equilibrium analysis, may lead to underestimation of economic impacts due to its optimizing behavior features	Potential effects of catastrophic earthquakes on regional economies and linked to lifeline analysis
Growth frameworks	Interrelationship between growth, development and disaster risk	Longer term framework for tracing disaster impacts	No short term dynamics	Impact on macroeconomic productivity, aggregate effects in developing countries exposed to disaster risk

Source: Adapted and extended based on Okuyama 2009

Few studies have applied economic growth models to the assessment of disaster impacts, one reason being a lack of robust information on capital stock at risk. Economic growth frameworks can be useful in assessing the longer term interrelationship between economic growth, development and disaster risk. The models have a long time horizon and thus do not account for short-term dynamics. Growth modeling has been used to challenge the *Creative Destruction Hypothesis* (see Cavallo and Noy 2009) that suggests that disasters help to "update" (i.e. destroy) inefficient capital stock and, given investments in more efficient capital, lead to increased growth in the longer term. Hallegatte and Dumas (2008) find that effects on productivity may decrease or increase the costs of disasters, but do not have the potential to lead to higher growth post-disaster. Growth modeling has also been used to study the longer-term consequences of disaster events and risk in developing countries highly exposed to disaster risk as a function of the availability of domestic savings and inflow of external savings (Freeman et al. 2002; Mechler et al. 2006; Hochrainer 2006).

6.3 Modeling Disaster Risk Explicitly

Given the *fat tailed* feature of natural disaster risk, disaster risk modeling ideally derives probabilities and impacts for entire loss distributions (so called *loss exceedance distributions*). However, most macroeconomic modeling analyses have focused on reanalyzing one observed event, and only a few have aimed at representing extremes embedded in a risk-based, forward-looking framework (Freeman et al. 2002; Mechler 2004; Hochrainer 2006; Hallegatte and Ghil 2008). Some studies have represented risk in terms of a deterministic estimate (such as a 100-year event) or averages (expected annual loss). Overall, this does not lend itself to a forward-looking and comprehensive analysis of risk and may lead to a serious underestimation of the potential consequences of natural disasters, which by *nature* are low-probability, high-impact events.

6.4 Contributions in this Volume

Probabilistic analysis of macroeconomic risk forms the backbone of the discussions in the three chapters of this section. What makes these studies unique is their explicit focus on introducing risk into economic modeling frameworks.

Chapter 7 (*Catastrophe Risk and Economic Growth*) by Yuri Ermoliev and Tatiana Ermolieva analyses the effects of catastrophes on economic growth and stagnation using a stylized Harrod-Domar growth model (Ermoliev et al. 2012). They conclude that any short-term incremental analysis of economic growth underestimates the impacts of rare catastrophic shocks. Disasters shocks (also small-scale) can persist in time and lead to traps and thresholds, which trigger stagnation and shrinking even in developed economies. The study suggests that the stabilization of regional growth must rely on a proper combination of structural and financial ex-ante (anticipative) risk reduction and ex-post (adaptive) risk sharing and transfer options, such as, e.g. purchasing catastrophe bonds, credit and insurance. The need for the co-existence of both options is demonstrated in other chapters of the book (e.g. Chaps. 2 and 5). This chapter illustrates that ex-ante and ex-post policies for dealing with catastrophic shocks are not substitutable and must be analyzed jointly as complementary approaches for handling risk.

Chapter 8 (*Modeling Macro Scale Risk: The CATSIM Model*) by Stefan Hochrainer-Stigler, Reinhard Mechler and Georg Pflug presents the IIASA CATSIM (CATastrophe SIMulation) model, which is a risk-based growth model framework for evaluating economic disaster impacts and responses (Hochrainer-Stigler et al. 2012). Based on stochastic simulation of disaster risks in a specified country, it examines the ability of the government and private sector to finance relief and recovery. The model can be used for supporting policy planning processes for the allocation of resources between ex-ante spending on disaster risk management (such as prevention, national reserve funds, sovereign insurance) and ex-post spending on relief and reconstruction. This chapter sets out the model, which is applied to the cases of Nepal (Chap. 9, see next paragraph) and Hungary/ Tisza (Chap. 16) in this volume.

Chapter 9 (*Managing Indirect Economic Consequences of Disaster Risk: The Case of Nepal*) by Reinhard Mechler, Stefan Hochrainer-Stigler and Kazuyoshi Nakano applies IIASA's CATSIM model to the case of Nepal in order to quantify country level direct disaster risk as well as the corresponding indirect effects using growth modeling and input-output analysis (Mechler et al. 2012). The authors find that the economic and fiscal risks posed by natural disasters in Nepal are large and potentially long lasting, particularly when they are triggered by earthquake risk. Given these results, the authors suggest that there is a clear case for considering risk in economic and fiscal planning processes in Nepal and similar heavily disaster exposed countries.

6.5   Concluding Remarks

This introductory chapter reviewed methodologies and evidence on the macroeconomic effects caused by disasters. The discussion regarding the different modeling techniques and their advantages and downsides identifies one key challenge, which is the difficulty of explicitly accounting for disaster risk using probabilistic representations of disaster occurrence severity and impacts on the economy. This is the challenge forming the backbone of this section and is addressed in detail by the three chapter contributions.

References

Albala-Bertrand J (1993) Political economy of large natural disasters with special reference to developing countries. Clarendon, Oxford

Benson C, Clay E (2004) Understanding the economic and financial impacts of natural disasters. Disaster risk management series no. 4. The World Bank, Washington, DC

Boisvert R (1992) Indirect losses from a catastrophic earthquake and local, regional, and national interest. In: Federal Emergency Management Agency (ed) Indirect economic consequences of a catastrophic earthquake. Federal Emergency Management Agency, National Earthquake Hazards Reduction Program, Washington, DC

Brookshire D, McKee M (1992) Other indirect costs and losses from earthquakes: issues and estimation. In: Federal Emergency Management Agency (ed) Indirect economic consequences of a catastrophic earthquake. Federal Emergency Management Agency, National Earthquake Hazards Reduction Program, Washington, DC

Cavallo E, Noy I (2009) The economics of natural disasters: a survey, RES working papers 4649. Inter-American Development Bank, Washington, DC

Charveriat C (2000) Natural disasters in Latin America and the Caribbean: an overview of risk, RESworking papers 434. Inter-American Development Bank, Washington DC

Cho S, Gordon P, Moore J II, Richardson H, Shinozuka M, Chang S (2001) Integrating transportation network and regional economic models to estimate the costs of a large urban earthquake. J Reg Sci 41:39–65

Cochrane H, Chang S, Rose A (1997) Indirect economic losses. In: National Institute of Building Sciences (ed) Earthquake loss estimation methodology: HAZUS technical manual, vol 3. National Institute of Building Sciences, Washington, DC

Cole S (1995) Lifeline and livelihood: a social accounting matrix approach to calamity preparedness. J Conting Crisis Manag 3:228–240

Cole S (1998) Decision support for calamity preparedness: socioeconomic and interregional impacts. In: Shinozuka M, Rose A, Eguchi R (eds) Engineering and socioeconomic impacts of earthquakes. Multidisciplinary Center for Earthquake Engineering Research, Buffalo

Cole S (2004) Geohazards in social systems: an insurance matrix approach. In: Okuyama Y, Chang S (eds) Modeling spatial and economic impacts of disasters. Springer, New York

ECLAC (2003) Handbook for estimating the socio-economic and environmental effects of disaster. Economic Commission for Latin America and the Caribbean, Mexico City

Ellson R, Milliman J, Roberts R (1984) Measuring the regional economic effects of earthquakes and earthquake predictions. J Reg Sci 24(4):559–579

Ermoliev Y, Ermolieva T (2012) Catastrophe risk and economic growth, Chapter 7. In: Amendola A, Ermolieva T, Linnerooth-Bayer J, Mechler R (eds) Integrated catastrophic risk modeling. Supporting policy processes. Springer, Heidelberg

Freeman P, Martin L, Mechler R, Warner K with Hausman P. (2002) Catastrophes and development, integrating natural catastrophes into development planning, Disaster risk management working paper series no.4. World Bank, Washington, DC

Gordon P, Richardson H, Davis B (1998) Transport-related impacts of the Northridge earthquake. J Transp Stat 1:22–36

Hallegatte S, Dumas P (2008) Can natural disasters have positive consequences? Investigating the role of embodied technical change. Ecol Econ 68:777–786

Hallegatte S, Ghil M (2008) Natural disasters impacting a macroeconomic model with endogenous dynamics. Ecol Econ 68:582–592

Handmer J, Honda Y, Kundzewicz Z, Arnell A, Benito G, Hatfield J, Mohamed I, Peduzzi P, Wu S, Sherstyukov B, Takahashi K, Yan Z (2012) Changes in impacts of climate extremes: human systems and ecosystems. In: Field C, Barros V, Stocker T, Qin D, Dokken D, Ebi K, Mastrandrea M, Mach K, Plattner G-K, Allen S, Tignor M, Midgley P (eds) Intergovernmental panel on climate change special report on managing the risks of extreme events and disasters to advance climate change adaptation. Cambridge University Press, Cambridge/New York

Hochrainer S (2006) Macroeconomic risk management against natural disasters. German University Press, Wiesbaden

Hochrainer S (2009) Assessing macroeconomic impacts of natural disasters: are there any? Policy research working paper 4968. World Bank, Washington, DC

Hochrainer-Stigler S, Mechler R, Pflug G (2012) Modeling macro scale risk: the CATSIM model, Chapter 8. In: Amendola A, Ermolieva T, Linnerooth-Bayer J, Mechler R (eds) Integrated catastrophic risk modeling. Supporting policy processes. Springer, Heidelberg

Kellenberg D, Mobarak A (2008) Does rising income increase or decrease damage risk from natural disasters? J Urban Econ 63:788–802

Lal P, Mitchell T, Aldunce P, Auld H, Mechler R, Miyan A, Romano L, Zakaria S (2012) National systems for managing the risks from climate extremes and disasters. In: Field C, Barros V, Stocker T, Qin D, Dokken D, Ebi K, Mastrandrea M, Mach K, Plattner G-K, Allen S, Tignor M, Midgley P (eds) Intergovernmental panel on climate change special report on managing the risks of extreme events and disasters to advance climate change adaptation. Cambridge University Press, Cambridge/New York

Mechler R (2004) Natural disaster risk management and financing disaster losses in developing countries. Verlag für Versicherungswissenschaft, Karlsruhe

Mechler R, Linnerooth-Bayer J, Hochrainer S, Pflug G, Pflug G (2006) Assessing financial vulnerability and coping capacity: the IIASA CATSIM model. In: Birkmann J (ed) Measuring vulnerability and coping capacity to hazards of natural origin. Concepts and methods. United Nations University Press, Tokyo

Mechler R, Hochrainer-Stigler S, Nakano K (2012) Managing indirect economic consequences of disaster risk: the case of Nepal, Chapter 9. In: Amendola A, Ermolieva T, Linnerooth-Bayer J, Mechler R (eds) Integrated catastrophic risk modeling. Supporting policy processes. Springer, Heidelberg

Noy I (2009) The macroeconomic consequences of disasters. J Develop Econ 88(2):221–231

Okuyama Y (2009) Critical review of methodologies on disaster impacts estimation, Background paper for World Bank/UN report unnatural disasters. World Bank, Washington, DC

Otero R, Marti R (1995) The impacts of natural disasters on developing economies: implications for the international development and disaster community. In: Munasinghe M, Clarke C (eds) Disaster prevention for sustainable development: economic and policy issues. World Bank, Washington, DC

Raddatz C (2007) Are external shocks responsible for the instability of output in low-income countries? J Develop Econ 84:155–187

Rose A (1981) Utility lifelines and economic activity in the context of earthquakes. In: Isenberg J (ed) Social and economic impact of earthquakes on utility lifelines. American Society of Civil Engineers, New York

Rose A (2004) Economic principles, issues, and research priorities in hazard loss estimation. In: Okuyama Y, Chang S (eds) Modeling spatial and economic impacts of disasters. Springer, New York

Rose A, Benavides J (1998) Regional economic impacts. In: Shinozuka M, Rose A, Eguchi R (eds) Engineering and socioeconomic impacts of earthquakes. Multidisciplinary Center for Earthquake Engineering Research, Buffalo

Rose A, Guha G (2004) Computable general equilibrium modeling of electric utility lifeline losses from earthquakes. In: Okuyama Y, Chang S (eds) Modeling spatial and economic impacts of disasters. Springer, New York

Rose A, Liao S (2005) Modeling regional economic resilience to disasters: a computable general equilibrium analysis of water service disruptions. J Reg Sci 45:75–112

Rose A, Benavides J, Chang S, Szczesniak P, Lim D (1997) The regional economic impact of an earthquake: direct and indirect effects of electricity lifeline disruptions. J Reg Sci 37(3): 437–458

Skidmore M, Toya H (2002) Do natural disasters promote long-term growth? Econ Inq 40(4): 664–687

Sohn J, Hewings G, Kim T, Lee J, Jang S (2004) Analysis of economic impacts of an earthquake on a transportation network. In: Okuyama Y, Chang S (eds) Modeling spatial and economic impacts of disasters. Springer, New York

Chapter 7
Economic Growth Under Catastrophes

Yuri Ermoliev and Tatiana Ermolieva

Abstract The chapter analyzes effects of catastrophes on economic growth and stagnation. The economy is a complex system constantly facing shocks and changes with possible catastrophic impacts. A shock is understood as an event removing from the economy a part of the capital. We show that even in the case of well-behaving economies defined by the Harrod-Domar model, persistent in time shocks implicitly modify the economy and may lead to various traps and thresholds triggering stagnation and shrinking. The stabilization of the growth must then rely on ex-ante risk reduction and risk transfer options, such as hazard mitigation and the purchase of catastrophic insurance, as well as ex-post borrowing. The coexistence of ex-ante (risk averse) and ex-post (risk prone) options in the proposed model generates a strong risk aversion even in the case of linear utility functions. In contrast to the traditional expected utility theory, it assesses and explains trade-offs and benefits of ex-ante and ex-post management options.

Keywords Catastrophic risks • Economic growth under shock • Growth stabilization • Ex-ante and ex-post measures • Risk aversion • Two-stage stochastic optimization

Y. Ermoliev
International Institute for Applied Systems Analysis (IIASA),
Schlossplatz 1, A-2361 Laxenburg, Austria

T. Ermolieva (✉)
Ecosystems, Services and Management (ESM) Program, International Institute for Applied Systems Analysis (IIASA), Schlossplatz 1, A-2361 Laxenburg, Austria
e-mail: ermol@iiasa.ac.at

A. Amendola et al. (eds.), *Integrated Catastrophe Risk Modeling: Supporting Policy Processes*, Advances in Natural and Technological Hazards Research 32,
DOI 10.1007/978-94-007-2226-2_7, © Springer Science+Business Media Dordrecht 2013

7.1 Introduction[1]

As an alarming tendency of current global changes, losses from human-made and natural catastrophes are rapidly increasing (Munich Re 2011; IPCC 2011). In addition to destruction of human lives, infrastructure and assets, they affect consumption, savings and investments. The direct economic costs are split approximately equally between the developed and developing countries, but especially sensitive are low-income countries. In the standard economic theory there is no special problem of even catastrophic risks (Arrow 1996). It is assumed that all economic agents know all possible shocks (states of the world), i.e., they know when, how often, and what may happen to each of them. Therefore, they can easily organize "markets", where everyone insures everyone by pooling resources available in any state of the entire society, i.e., a catastrophe becomes small on the scale of the world. In reality this pool does not exist, which calls for more realistic models with explicit representation of uncertainties and associated risks. Especially important are models that explain connections between poverty, stagnation, and shocks. One of the reasons which could cause low growth (Easterly 1994; Ray 1998; Linnerooth-Bayer et al. 2011) is the low saving rates typically observed in low-income countries: economies where the majority of citizens have incomes close to minimum of subsistence level are unlikely to have a high rate of savings. But the low saving rate is not sufficient to explain within conventional models why the sustained growth may not "take-off".

This chapter analyzes effects of catastrophes or shocks on economic growth and stagnation. A shock is understood as an event destroying a part of capital stock. For example, shocks may be due to natural and human-made disasters or the flight of capital from the country. We show that even in the case of such well-behaved economies as the economies defined by the Harrod-Domar model (Harrod 1939), persistent in time shocks implicitly modify the production function of the economy and may lead to various traps and thresholds triggering stagnation and shrinking.

The stabilization of growth may then rely on various defensive mechanisms such as post-event borrowing, ex-ante loss reduction measures and mechanisms for loss spreading through insurance and financial markets. As the economy grows, crossing certain instability levels, the assistance for growth can be reduced or completely disappear. Section 7.2 analyzes effects of shocks on sustained growth. In Sect. 7.3 we illustrate that ex-ante and ex-post policies for dealing with shocks are not substitutes and must be analyzed jointly as complementary decisions. The proposed two-stage dynamic stochastic programming model incorporates both risk-averse (ex-ante) and risk-prone (ex-post) decisions, and, in contrast to the traditional models of the expected utility theory, provides more realistic decision making framework. According to this framework, only some decisions are made ex-ante, whereas other options are kept open until more information about shocks becomes

---

[1] This chapter is based on the paper "Economic Growth Under Shocks: (Ermoliev 2006) Path Dependencies and Stabilization" The paper is reprinted with permission of the publisher.

available and, hence, can be better utilized in ex-post decisions. In proposed models a strong risk aversion occurs even for linear utility functions. Section 7.4 examines the convergence of an economy under shocks to a path of sustained growth. Shocks implicitly modify the economy, and the convergence becomes impossible without appropriate growth efforts to by-pass various traps and thresholds. Section 7.5 concludes.

7.2 Sustained Economic Growth

To better understand the effects of catastrophes on economic growth let us consider a well-behaved economy, characterized by the production function with two factors: "capital" and "labor", $Y = F(K, L)$, and constant returns to scale, $Y = LF$ $(K/L, 1)$. Therefore, we can characterize the economy in terms of capital to labor ratio, $k = K/L$, and output to labor ratio, $y = Y/L, y = f(k) := F(k, 1)$. Assume that output Y is subdivided into consumption and savings, and savings are equal to investments. The growth is driven by the accumulation of capital through investments

$$\frac{dK}{dt} = I - \delta K, K(0) = K_0, t > 0, \tag{7.1}$$

where δ, $0 < \delta < 1$, is the capital depreciation rate. Assume further that the investments $I(t)$ are simply a fraction s, $0 < s < 1$, of the output, i.e., $I(t) = sY(t)$, and γ is an exponential growth rate of the population, $\frac{d}{dt}\ln L = \gamma$. We can then rewrite (7.1) in variables k:

$$\frac{dk}{dt} = sf(k) - (\gamma + \delta)k, k(0) = k_0, t > 0, \tag{7.2}$$

or

$$\frac{d}{dt}\ln k = s\frac{f(k)}{k} - \gamma - \delta. \tag{7.3}$$

If the output to capital ratio is constant θ, i.e., $y/k = f(k)/k = \theta$, then it leads us to the very influential Harrod-Domar (Ray 1998; Solow 1997) model with constant exponential rate of growth

$$\frac{d}{dt}\ln k = s\theta - \gamma - \delta. \tag{7.4}$$

According to (7.4), the rate of growth is determined jointly by the saving rates and the productivity of capital θ, that is, the inverse of the capital-output ratio. Since the growth in real output $\frac{d}{dt} \ln y(t)$ is the same as the growth in the capital stock $\frac{d}{dt} \ln k(t)$, then from (7.4) it follows that the exponential growth is defined by linear function

$$\ln y = \ln k = k_0 + (s\theta - \gamma - \delta)t, t > 0. \tag{7.5}$$

The economy is a complex system constantly facing shocks and changes. A catastrophe is one of such shocks. We can model shocks similarly to the depletion of capital reducing the rate of growth $s\theta - (n + \delta)$ to a random level $s\theta - (n + \delta - v)$, $v = v(t, k, \omega)$, $0 \leqslant v < 1$, where $v(t, k, \omega)$ denotes impacts of the shock ω at the current level of $k(t)$. In the case of catastrophes shocks are rare events which may occur at some random time moments $T_0, T_1, T_2, \ldots, T_0 = 0$, $v(0, \tilde{\cdot} \cdot \tilde{\cdot}) = 0$. In this conceptual model the random intensity v depends on the aggregate current level $k(t)$ of the capital and the random shock ω. In realistic versions of the model $v(t, k, \omega)$ must depend on the geographical distribution of $k(t)$ and ω as well as on other country-specific sources of vulnerability. The accumulation of investments in specific risk prone regions and sectors of the economy can make a significant difference for the probability distribution of v (t, k, ω). The existence of an unreliable system of banks can further magnify impacts of shocks from natural disasters by provoking the capital flight from the country. These details provide necessary information for the design of appropriate policies. Shocks, in general, transform linear function (7.5) into highly nonlinear (discontinuous) random function

$$\ln y(t) = k_0 + (s\theta - \gamma - \delta)t - V(t), V(t) = \sum_{t=1}^{N(t)} v_i, \tag{7.6}$$

where $N(t)$ is the number of shocks in the interval $(0, t]$, and v_i is the size of these shocks. Assume that random variables v_1, v_2, $\ldots$ are independent, identically distributed, and they are independent of time between shocks $\tau_i = T_i - T_{i-1}$.

Remark: Informally, (7.6) specifies that if shocks represent a pure jumping process with isolated jump times $T_1, T_2, \ldots$, then between jumps the exponential growth $G(t) := \ln y(t)$ satisfies the differential equation

$$\frac{d}{dt} G(t) = (s\theta - \delta - \gamma), T_{i-1} \leqslant t < T_i.$$

The function $G(t)$ is right-continuous and the value of $G(T_{i-})$ just before the jump is defined as $G(T_{i-}) := \lim_{t \uparrow T_i} G(t)$. After a jump, motion restarts as before according to the same differential equation. If $G_i := G(T_i)$, $\Delta G_i := G(T_i) - G(T_{i-1}) = g\tau_i$ $-v_i$, $\Delta G_0 = 0$, $g := s\theta - \gamma - \delta$, then $\{\Delta G_i\}_{i=0}^{\infty}$ is a sequence of independent identically distributed random variables,

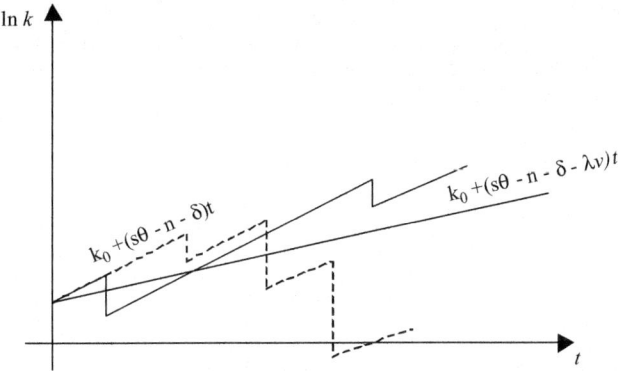

Fig. 7.1 Expected and real growth rates

$$G(t) = k_0 + g(t - T_{N(t)}) + \sum_{i=0}^{N(t)} G_i.$$

This remark connects the economic growth processes with the so-called piecewise-deterministic Markov processes (Davis 1984) and the renewal processes. If time between the shocks has stationary distribution with mathematical expectation λ, then the expected exponential growth is still characterized by a linear in t function (see Fig. 7.1), $E \ln y(t) = k_0 + (s\theta - \gamma - \delta - \lambda\mu)t$, and it is positive when the rate of investments exceeds the average rate of losses, $s\theta - \gamma - \delta - \lambda\mu > 0$. From the strong law of large numbers it follows that

$$\frac{\ln y(t, \omega)}{t} \rightarrow s\theta - \gamma - \delta - \lambda\mu$$

with probability 1. It means that for each possible random growth path, denote it as $\omega = (T_1, v_1, T_2, v_2, \ldots)$, the sustained growth "takes-off" only after a random time $T(\omega)$, i.e., $\ln y(t, \omega) \approx (s\theta - \gamma - \delta - \lambda\mu)t > 0$, $\forall t > T(\omega)$. On the way to the sustained growth for $t < T(\omega)$ the economy may stagnate and even shrink. In other words, for a given t there may exist a positive probability that accumulated random losses exceed accumulated investments

$$k_0 + (s\theta - \gamma - \delta)t - V(t) \leq 0. \tag{7.7}$$

Let us note that the analysis of the event (7.7) is similar to the analysis of the so-called ruin of insurers (Dupačovlá and Bertocchi 1995; Embrechts et al. 2000). An important indicator of stability is the probability that the accumulation of growth between shocks exceeds possible losses, i.e.,

$$(s\theta - \delta - \gamma)\tau - v(t, \tau, \omega) \geq g\tau, \tag{7.8}$$

where $g > 0$ is a given target of the growth rate and $v(t, \tau)$ are losses associated with the first catastrophe which occurs at time $t + \tau$ after t. The economy might shrink or stagnate at t if there is positive probability of the event

$$(s\theta - \delta - \gamma)\tau - v(t, \tau, \omega) \leq 0. \tag{7.9}$$

Since positive probability of (7.7) is a consequence of (7.8), (7.9), these indicators can be used to protect the growth of the economy.

7.3   Programming Adjustment of Growth: Ex-Ante and Ex-Post Strategies

The common practice to deal with shocks is to use such ex-post measures as borrowing, foreign aid, and the diversion of investments committed to other needs. This section illustrates the importance of both ex-ante and ex-post mechanisms in order to be prepared for shocks before they occur, as well as to have enough flexibility to react more effectively to revealed situations.

The Harrod-Domar model is a building block in the growth programming approach to economic development (Khan et al. 1990). According to (7.4), the overall rate of growth in the interval $[t, t + \Delta t]$ is approximately defined as

$$[Y(t + \Delta t) - Y(t)]/\Delta t Y(t)$$
$$= [K(t + \Delta t) - K(t)]/\Delta t K(t) \approx s\theta - \gamma - \delta. \tag{7.10}$$

This equation links the growth rate of the economy to two fundamental variables: the ability of the economy to save and the productivity of the capital θ. By increasing the rate of savings s or capital productivity θ, it would be possible to accelerate the rate of growth. This creates the basis for programming of the growth rate, i.e., the design of policies which provide such levels of parameters s, θ, γ, δ, that guarantee a given level of g. A "gap" between available investments and the investment required to achieve the output growth target, e.g., due to "unforeseen" shock, provides the information for the borrowing needed to cover this "gap".

Equation (7.10) has rather conceptual, symbolic character. In realistic models values of all parameters s, θ, γ, δ are defined by various components. For example, the distribution of incomes among the population and investments among various sectors of the economy and geographical regions may play a critical role for the stability of the growth. If some of the parameters s, θ, γ, δ or their components are fixed, then missing values can be defined from (7.10). As it was pointed out by Ray (1998, p. 58), such "if-then" analysis in many cases does not make sense, since the parameters that are used to predict growth rate may themselves be affected by the growth processes. Besides this, critical problems arise with the naïve short-term sequential adjustments of growth rates in the presence of shocks. The analysis of the growth rate at time t only for the next interval $[t, t + \Delta t]$ may not provide a good idea to develop preparedness and loss reduction measures. The occurrence of a rare catastrophe in each interval of length t is usually evaluated by negligibly small probability $\lambda \Delta t$, but the probability of a catastrophe in an interval $[0, T]$ $1 - (1 - \lambda \Delta t)^{T/\Delta t} \approx 1 - e^{\lambda T}$ is dramatically increasing with T. Therefore, the economy must take a long-term perspective in order to develop appropriate measures against rare ("unforeseen" otherwise) shocks.

It is important that all possible ex-ante and ex-post decisions are not evaluated in a sequential, "one-by-one" manner, since the most important is the synergy of them. Thus a certain level of mitigation measures increasing, say, resistance of buildings to earthquakes, may essentially increase their insurability. On the other hand, the existence of insurance may enforce the mitigation measures through appropriate reductions in prices of insurance contracts.

The coexistence of risk prone and risk averse decisions within the same model can be viewed as a rather flexible decision making framework when we commit ourselves "ex-ante" only to a part of possible decisions and, at the same time, we keep other options open until more information becomes available and can be effectively utilized by appropriate ex-post options. As we can see further, this type of models, the so-called two-stage stochastic optimization models, in contrast to the standard model of the expected utility theory, produces strong risk aversion even for linear utility functions.

Consider the following important growth adjustment problem where a part of the growth rate is used to protect a given target value of the sustained growth path. According to the model (7.6) the value of the growth rate $g = s\theta - \gamma - \delta$ may be interrupted by a shock ω. In order to protect g it is possible to mobilize ex-post internal resources by reducing consumption, increasing inflation, by external aid and borrowing. Let us denote this type of decisions by $y(\omega)$, where ω indicates that ex-post decisions are made on the basis of information about observed shock ω. Besides, the society can also protect, reduce or spread losses through various ex-ante measures. Let us denote this type of decisions by x. They are chosen before the occurrence of the shock ω and, therefore, they do not depend on ω. In general, x and $y(\omega)$ are vectors with components characterizing different options, for example, the first component of x may correspond to a certain measure for the reinforcement of buildings, second – to an insurance contract, another pair of components – to a weather related bond characterized by two decision variables specifying its "trigger" and the "cap". The adjustment of the growth rate is very simple from the methodological point of view when among policy options there are only traditional ex-post strategies, for example, borrowing. Each of them is evaluated only against a revealed situation. The fundamental challenge with the presence of ex-ante strategies is that they must be evaluated against all possible shocks in order to be robust against them. Since this requires strong computational approaches, we consider a discrete time version of (7.6). Let as before $G(t) = \ln y(t) = \ln k(t)$. Consider a time interval $[0, T]$ subdivided into N subintervals of length $t = T/N$, $t_s = t_{s-1} + \Delta t$, $s = 1, 2, \ldots , N$, $t_0 = 0$. We assume that a shock ω_t at time $t = t_s$ comprises all events in $[t_s - \Delta, t_s]$.

In the same manner losses $v_t, t = t_s, s = 1, 2, \ldots, N$ from ω_t comprise aggregate losses from all events in $[t_s - \Delta, t_s]$. For the simplicity of notations we assume that v_t may be equal 0 and $\Delta t = 1$, i.e., we simply consider time moments $t = 0, 1, \ldots, N$. We also do not indicate implicitly the dependence of $G(t), v_t, t = 1, \ldots, N$, on the current level of per capita capital $k(t)$. Variables $G(t), v_t$ in general depend on the history of shocks $\omega^t = (\omega_1, \ldots, \omega_t)$. In the following we use the notations G (t, ω), v_t (ω) to indicate this dependence, where $\omega = (\omega_1, \ldots, \omega_N)$. A vector x of ex-ante decisions chosen at $t = 0$ transforms v_t (ω) into v_t (x, ω), whereas $G(t)$ depends on both ex-ante and ex-post decisions G $(t, x, y^t (\omega), \omega), y^t(\omega) = (y_1(\omega), \ldots, y_t(\omega))$. By using these notations we can write the discrete time version of (7.6) in the form

$$G(t, x, y^t, \omega) = G(t - 1, x, y^{t-1}, \omega) + g - v_t(x, \omega) + y_t - (1 + \beta_{t-1})y_{t-1}, \quad (7.11)$$

where $t = 1, \ldots, N, x \in X \subset R^n$ and X is a compact set of feasible ex-ante decisions, y_t is the amount of borrowing (credits) in $[t - 1, t], y^t = (y_0, y_1, \ldots, y_t)$. In this model we assume that credits at t are paid at $t + 1$ with the interest rate β_t, which may require borrowing at $t + 1$, and so on.

The problem is to find a combination $(x, y(\omega))$ of ex-ante and ex-post decisions maximizing the expected accumulated growth rate in $[0, T]$, i.e., the expectation function

$$E\left[G(T, x, y^T(\omega), \omega) - (1 + \beta_T)y_T(\omega)\right], \quad (7.12)$$

ensuring a given target growth rate q:

$$G(t, x, y^t, \omega) \geq qt, t = 1, 2, \ldots, T, \text{ for all } \omega. \quad (7.13)$$

In general, q may also be a decision variable, but here we assume that q is fixed. We can subtract q from g in (7.11) and transform (7.13) into requirements

$$G(t, x, y^t, \omega) \geq 0, t = 1, 2, \ldots, T, \text{ for all } \omega. \quad (7.14)$$

Therefore, in the following we consider (7.14) instead of (7.13), assuming $g = s\theta - \gamma - \delta - q$. In this model a target value q of the growth rate is achieved by sacrificing a portion of the economic growth for defensive ex-ante and ex-post measures. Let us note that $G(t, x, y^t(\omega), \omega)$ is a linear in y^t (ω) function and it may also be a linear function in x, i.e., a risk neutral utility function. The important feature of the model (7.11), (7.12), (7.14) is that it incorporates both ex-ante x and ex-post y (ω) decisions, which generates a strong risk aversion with respect to x. Indeed, $G(t, x, y^t(\omega), \omega)$ can be written in the form

$$G(t, x, y^t(\omega), \omega) = \overline{G}_t(x, \omega) - \sum_{s=1}^{t-1} \beta_s y_s + y_t,$$

where $\overline{G}_t(t,\omega) = k_0 + gt - \sum_{s=1}^{t-1} v_s(x,\omega)$. From this it follows that y_t satisfies equation

$$y_t = \max\left\{0, -\overline{G}_t + \sum_{s-1}^{t-1} \beta_s y_s\right\}. \tag{7.15}$$

Therefore, the problem (7.11), (7.12), (7.14) can be rewritten as the maximization of the expectation function

$$F(x) = Ef(x,\omega), \tag{7.16}$$

which depends only on ex-ante decisions $x \in X$, where

$$f(x,\omega) = \overline{G}_T + \beta_1 \min\{0,\overline{G}_1\} + \beta_2 \min[0,\overline{G}_2 + \beta_1 \min\{0,\overline{R}_1\}$$
$$+ \beta_3 \min\{0,\overline{R}_3 + \beta_1 \min\{0,\overline{R}_1\} + \beta_2 \min\{0,\overline{R}_2\}\}] + \cdots$$

In other words, the problem (7.11), (7.12), (7.14) is equivalent to the maximization of the nonlinear in x expectation function (7.16) and the calculation of ex-post decisions according to the recursive equation (7.15). If the probability distribution of v_t for some $t = 1, \ldots, N$ has a continuous density function, then the expectation function (7.16) is a strictly concave function since $f(x, \omega)$ is formed by using the operation min of linear in x functions. The described above implicit nonlinear character of the problem (7.11), (7.12), (7.14) with respect to x is due to the nature of ex-post decision $y(\omega)$. It is made on the basis of revealed shock ω and given x, i.e., $y(\omega)$ depends implicitly on x. This is a general feature of the two-stage stochastic optimization problems (Ermoliev and Wets 1988). The model (7.11), (7.12), (7.14) is a dynamic two-stage stochastic optimization problem, which can be solved by specific stochastic optimization procedures.

The need for coexistence of ex-ante and ex-post decisions becomes more evident from the following simple situation. Assume that there are only two time intervals: "now", $t = 0$, and "future", $t = 1$. Ex-ante decision x is made at $t - 0$, whereas ex-post decision $y(\omega)$ is made at $t = 1$ on the basis of information about impacts $v(\omega)$ of shock ω and the decision x. The decision x protects all losses $v(\omega)$ below the level x at the cost cx, i.e., $v(\omega)$ is transformed into the function

$$v(x,\omega) = \begin{cases} v(\omega) - x, & \text{if } x \le v(\omega), \\ 0, & \text{otherwise.} \end{cases}$$

The ex-post decision, borrowing $y(\omega)$, protects the economy from losses $v(\omega) - x$, $x \le v(\omega)$, at the cost $(1 + \beta(\omega))y(\omega)$. The problem is to minimize expected value

$$cx + E(1 + \beta(\omega))y(\omega), x \ge 0,$$

where $y(\omega) \geq v(\omega) - x$ for $x \leq v(\omega)$. It is clear that $y(\omega) = \max\{0, v(\omega) - x\}$, i.e., $y(\omega)$ depends on (x, ω). A solution of this problem may be $x = 0$, which means, totally rely on the borrowing. The analogue of the maximization problem (7.16) in this simple case is the minimization of the function

$$F(x) = cx + E(1 + \beta(\omega)) \max\{0, v(\omega) - x\}$$

$$= cx + \int_x^{\bar{v}} (v - x)\varphi(v)dv + \int_x^{\bar{v}} \beta(v)(v - x)\varphi(v)dv,$$

for $x \geq 0$. Here we assume that the probability distribution of v has the support $[0, \bar{v}]$, $v(\omega) \in [0, \bar{v}]$, with a continuous density function $\varphi(v)$. From this it follows that $F(x)$ has the continuous derivative $F'(x)$ and, as it is easy to verify,

$$F'(x) = c - \int_x^{\bar{v}} \varphi(v)dv - \int_x^{\bar{v}} \beta(v)\varphi(v)dv.$$

The function $F'(x)$ is monotonically increasing for $x \to \bar{v}$. Therefore, if $c < \int_{\bar{v}}^{\bar{v}} \varphi(v)dv + \int_0^{\bar{v}} \beta(v)\varphi(v)dv$, or $c < 1 + E\beta(\omega)$, then there is a positive value $x = x^*$, $x^* \neq \bar{v}$, such that $F'(x^*) = 0$. Here $x^* = \bar{v}$ is excluded because $c > 0$. If $\beta(\omega)$ does not depend on ω, $\beta(\omega) = \beta$, then x^* is a quantile of $v(\omega)$ satisfying the equation

$$P[v > x] = \frac{c - 1}{\beta}.$$

Therefore, the minimization of the linear expected disutility function with different coefficients $(c, 1 + \beta)$ does not lead to the dominance of the preferable on average ex-ante solution $(c < 1 + \beta)$, i.e., both ex-ante and ex-post solutions coexist.

Let us note that the notion of ex-ante solution x reflects the fact that this type of solutions cannot be chosen differently after each new observation of shock ω, for example, the decision x may be a height of a water wall protecting the economy of a region against floods below level x. Generally, insurance contracts or weather-related bonds also cannot be issued as a function of every possible shock ω. This restricts the use of recursive equations (Sargent 1978) of the optimal control theory dealing only with the "feedback" control strategies.

7.4   Convergence, Traps and Thresholds

Section 7.2 shows that even for well-defined economies with the growth rate on average exceeding the losses from shocks the sustained growth in the presence of shocks takes off only in the long run. This conclusion radically changes when parameters s, θ, γ and losses v_i of the model are affected by the growth, i.e., they depend on the level of developments characterized in the model by per capita capital k or per capita income y. The rate of savings s may critically depend on the overall level y and its distribution in the society. Obviously, at low level of income, rates of savings are small. In this case a shock may further reduce them even to negative values (borrowing). As the economy grows there is increasing room for savings (Ray 1998, p. 59), but this does not necessarily mean that savings will grow steadily. To endogenize the dependence of s on y requires certain assumptions, the most important of them is the assumption of utility maximizing consumers. We do not follow this approach in the chapter. Instead, let us make a rather optimistic assumption that s is a linear function of y, $s = \alpha y = \alpha f(k) := s(k)$. A shock ω is modeled as a reduction of k to a random level $k(\omega)$ and it can be characterized by a probability distribution with the support in the interval $[0,k]$. Assume it has a rather smooth density function (k,v), i.e., the probability of losses in the interval $[v, v + dv]$, $k(\omega) \, \varepsilon [v, v + dv]$, is $\Psi(k,v)dv$. Then the expected value of random $s(k(\omega))$ is $s(k) = \int_0^k f(v)(k,v)dv$. It is easy to see that the second derivative $s''(k)$ involves the derivative $''_v(k, \cdot)$, which may be positive and negative at different points v. Therefore, $s''(k)$ may also have oscillating character. A similar situation occurs with the frequency λ and the expected losses μ. An exact evaluation of dependencies of λ and μ on k requires the use of geographically explicit catastrophe models, data on the vulnerability of engineering constructions and values at risk (see, for example, discussion in Ermoliev et al. 1998; Walker 1997). Assume that v_t is the sum of different fractions $k_1 = \beta_1 k, k_2 = \beta_2 k, \beta_1, \beta_2 \in (0, 1)$ of per capita $k(t)$, and they are allocated in two risk prone regions. Assuming the independence of shocks and losses in these regions, the expected losses $\mu(k)$ would have the following form

$$\mu(k) - \int_0^{k_1} v\varphi_1(k,v)dv + \int v\varphi_2(k_2,v)dv, k_1 = \beta_1 k, k_2 = \beta_2 k.$$

This function, similar to $s(k)$, may also have oscillating character. Figure 7.2 exhibits possible functions of $s(k)$, $\mu(k)$. Assume that the dynamics of growth under shocks can be written in a form similar to equation (4)

$$\frac{d}{dt} \ln k = s(k,\omega)\theta - \gamma - \delta - v(k,\omega), \qquad (7.17)$$

i.e., here we ignore purely jumping character of shocks by assuming that they affect the economy constantly.

The analysis of purely jumping processes requires a considerable extension of the chapter. From equation (7.4) follows that from the current level of

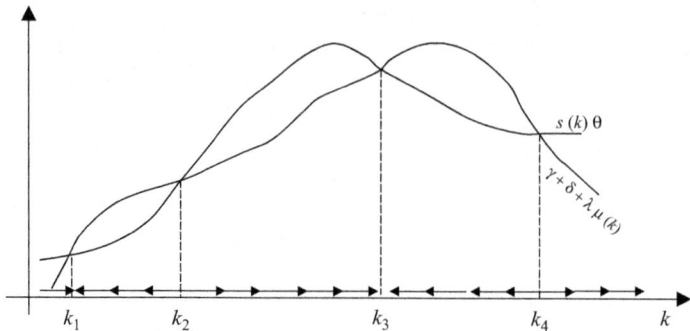

Fig. 7.2 Traps and thresholds

developments $k(t)$ the per capita capital moves in $[t, t + \Delta t]$ to the next random level $k(t + \Delta t)$ such that

$$\frac{E[k(t + \Delta t)|k(t)] - k(t)}{\Delta t} \approx k(t)(s(k)\theta - \gamma - \delta - \lambda\mu(k)),$$

where $E[\cdot|k(t)]$ denotes the conditional expectation of $k(t + \Delta t)$ for a given $k(t)$ Starting with a very low level of per capita income from the interval $[0, k_1]$, the expected rate of per capita income $g(k) = s(k)\theta - \gamma - \delta - \lambda\mu(k)$ is positive, the expected per capita will increase, i.e., the economy is expected to become richer. If the economy happens to be in the intervals (k_1, k_2), the expected per capita will decrease towards k_1, i.e., the state k_1 is a trap for the economy. If the economy is in the interval (k_1, k_3), the expected per capita income will increase over time till the next trap k_2. Intervals of developments from the trap k_1 to the threshold k_2, and from the trap k_3 to the threshold k_4 are critically important for the assistance in growth. If the economy is pushed up to cross the threshold, the economy would enter a path of sustained growth till the next trap. But these conclusions concern only expectations. As we discussed in Sect. 7.2, even in the case of constant positive expected per capita growth rate, the real random path of the economy enters the path of sustained growth only in the long run. In our case of s and μ dependent on k, it is possible to show that the real random path of $k(t)$ converges with probability 1 to traps under non-degenerating shocks, which allows the growth path to leave the thresholds. In other words, starting from the same initial condition k_0, the economy may end up at different more or less deep traps and stagnate within them thereafter. In general, this conclusion follows from the fundamental relations between the asymptotic properties of the path $k(t)$ defined by the stochastic equations (7.17), the deterministic equation

$$\frac{d\bar{k}}{dt} = s(\bar{k})\theta - \gamma - \delta - \lambda(\bar{k})\mu(\bar{k}), \qquad (7.18)$$

and the paths defined by the following finite-difference approximations. Let us subdivide the time interval $[0, \infty)$ into subintervals of the length $\rho_0, \rho_1, \ldots, \rho_s \to 0$, $\sum_{s=0}^{\infty} \rho_s = \infty$. Define points $t_1, t_2, \ldots$ as $t_{s+1} = t_s + \rho_s$, $t_0 = 0$, $s = 0, 1, \ldots$ and sequences $\{k(t_s)\}$, $\{\bar{k}(t_s)\}$, $s = 0, 1, \ldots$ according to the following finite-difference equations

$$\frac{k_{s+1} - k_s}{\rho_s} = k_s[s(k_s)\theta - \gamma - \delta - v(k_s, \omega)], s = 0, 1, \ldots, \qquad (7.19)$$

$$\frac{\bar{k}_{s+1} - \bar{k}_s}{\rho_s} = \bar{k}_s[s(\bar{k}_s, \omega)\theta - \gamma - \delta - \lambda(\bar{k}_s)\mu(\bar{k}_s)], s = 0, 1, \ldots, \qquad (7.20)$$

where $k_s = k(t_s)$, $\bar{k}_s = \bar{k}(t_s)$. Then it can be shown (Belenki and Volkonski 1974; Kushner and Clark 1978) that under natural assumptions the asymptotic properties of the equations (7.17), (7.18), (7.19) and (7.20) coincide.

7.5 Concluding Remarks

Any rational strategy for managing increasing vulnerability requires understanding of involved risks and losses. Now existing catastrophe models (Walker 1997) primarily deal with the estimation of direct losses. Indirect losses include costs of business interruptions, jumps of prices with booms in some sectors, e.g., constriction and depression in others. The destruction of capital stock results in losses of outputs and, hence, decreases wages, profits, savings and investments. This chapter shows that shocks have deeper indirect consequences - they implicitly modify the economy and may cause the stagnation and shrinking even in the case of the Harrod-Domar model of sustained growth.

In our conceptual model the increasing vulnerability is characterized by a probability distribution of shocks to capital with support in the interval $[0, k]$, which depends on the current level of per capita capital k. Formally this model covers the case when hazard mitigation measures, for example, a dam, may trigger new developments in flood-prone areas with a possibility of catastrophic losses from a failure of the dam. Of course, any realistic model of shocks must be based on dynamic catastrophe models with specific (for a given economy) patterns of catastrophes and vulnerability of the capital.

Generally, the costs of the catastrophes are borne by the central government from the national budget. This often "unexpectedly" diverts resources from other planned projects and requires "unplanned" additional borrowing. Associated costs are concerned with a transfer of shocks to other sectors of the economy, squeezing of planned investments and possible increase of risk premiums on total foreign debt. Ex-ante hazard preparedness and loss reduction measures may significantly reduce "unexpected" costs of post-shock responses. The proposed model incorporates both ex-ante and ex-post growth efforts and provides a unified decision-making framework to assess benefits of their synergy. Since the standard utility theory separates risk-averse (ex-ante) and risk-prone (ex-post) decisions, the proposed framework is built on more general ideas of stochastic optimization.

Any short term incremental analysis of economic growth underestimates the impacts of rare catastrophic shocks. To purchase catastrophe insurance or bond would be a good solution if the catastrophe occurs tomorrow. But it may also occur in 5, 20, or 100 years, i.e., within a time interval sufficient for implementing loss reduction measures. From the formal point of view the analysis of the long-term growth efforts is similar to the design of insurance portfolios in the presence of catastrophic risks (Ermoliev et al. 1998, 2000). In both cases the main problem concerns the protection of growth of certain accumulation processes such as risk reserves of insurers or outputs of the economy. The concentration only on ex-post options significantly simplifies the analysis, since such measures have to be optimal with respect to a particular observed shock. An ex-ante solution has to be optimal (robust), in a sense, against all possible shocks. The search of trade-offs between ex-ante and ex-post options is impossible just by ranking them. The presence of ex-ante options also makes impossible to use the so-called recursive equations of the conventional control theory (Sargent 1978). Major challenges are connected with analytically intractable structures of realistic accounting models for growth processes (MacKellar and Ermolieva 1999) and induced by shocks non-convexities of the resulting stochastic models. The most promising approach is to use the sophisticated computational techniques of the stochastic optimization. This is becoming increasingly important in financial planning (Zenios 1993; Ziemba and Mulvy 1998).

References

Arrow J (1996) The theory of risk-bearing: small and great risks. J Risk Uncertain 12:103–111

Belenki VZ, Volkonski VA (1974) Iterative methods in game theory and programming. Nauka, Moscow, pp 40–73 (in Russian)

Davis MHA (1984) Piecewise-deterministic Markov processes: a general class of Non-diffusion stochastic models. J R Stat Soc B 46:353–388

Dupačová J, Bertocchi M (1995) Management of bond portfolios via stochastic programming: post -optimality and sensitivity analysis. In: Doležal J, Fidler J (eds) System modeling and optimization. Proc. Of the 17-th IFIP TC7 conference, Prague. Chapmen & Hall, London, pp 574–582

Easterly W (1994) Economic stagnation, fixed factors, and policy thresholds. J Monet Econ 33:525–557

Embrechts P, Klueppelberg C, Mikosch T (2000) Modeling extremal events for insurance and finance: applications of mathematics, stochastic modeling and applied probability, vol 33. Springer, Heidelberg

Ermoliev Y, Wets R (1988) Numerical techniques of stochastic optimization. Computational mathematics. Springer, Berlin

Ermoliev Y, Ermolieva T, MacDonald G, Norkin V (1998) On the design of catastrophic risk portfolios. International Institute of Applied System Analysis, Interim report IR-98-056, Laxenburg, Austria

Ermoliev Y, Ermolieva T, MacDonald G, Norkin V, Amendola A (2000) A system approach to management of catastrophic risks. Eur J Oper Res 122:452–460

Ermoliev YM, Ermolieva TY, Norkin VI (2006), in MICRO MESO MACRO: Addressing Complex Systems Couplings; H. Liljenstroem, U. Svedin (eds), World Scientific Publishing Co. Pte. Ltd, Singapore, pp 289–302.

Harrod RF (1939) An essay in dynamic theory. Econ J 49(193):14–33

IPCC (2011) Summary for policymakers. In: Field CB, Barros V, Stocker TF, Qin D, Dokken D, Ebi KL, Mastrandrea MD, Mach KJ, Plattner G-K, Allen S, Tignor M, Midgley PM (eds) Intergovernmental panel on climate change special report on managing the risks of extreme events and disasters to advance climate change adaptation. Cambridge University Press, Cambridge

Khan MS, Montiel P, Haque NV (1990) Adjustments with growth: relating the analytical approaches of the IMF and the world bank. J Dev Econ 32:155–179

Kushner HJ, Clark DS (1978) Stochastic approximation for constrained and unconstrained systems. Springer, Berlin

Linnerooth-Bayer J, Mechler R, Hochrainer-Stigler S (2011) Insurance against losses from natural disasters in developing countries. Evidence, gaps and the way forward. J Integr Disaster Risk Manag 1(1):1–23

MacKellar L, Ermolieva T (1999) The IIASA multiregional economic-demographic model: algebraic structure. International Institute of Applied Systems Analysis, Interim report IR-99-007, Laxenburg, Austria

Munich Re (2011) Topics geo. Natural catastrophes 2010: analyses, assessments, positions. Munich Reinsurance Company, Munich. http://www.munichre.com/publications/302-06735_en.pdf

Ray D (1998) Development economics. Princeton University Press, Princeton

Sargent TJ (1978) Dynamic macroeconomic theory. Harvard University Press, Cambridge

Solow R (1997) Growth theory: an exposition. Clarendon, Oxford

Walker G (1997) Current developments in catastrophe modelling. In: Britton NR, Oliver J (eds) Financial risk management for natural catastrophes. Griffith University, Brisbane, pp 17–35

Zenios SA (1993) Financial optimization. Cambridge University Press, Cambridge

Ziemba WT, Mulvy J (1998) World Wide asset and liability modeling. Cambridge University Press, Cambridge

Chapter 8
Modeling Macro Scale Disaster Risk: The CATSIM Model

Stefan Hochrainer-Stigler, Reinhard Mechler, and Georg Pflug

Abstract Developing countries are placing increasing emphasis on improving their preparedness for and management of disaster risk. We discuss the CATSIM (CATastropheSIMulation) model developed at IIASA for assistance in such planning exercises. CATSIM represents a simple but *risk-based* economic framework for evaluating economic disaster impacts, and the costs and benefits of measures for reducing those impacts. CATSIM uses stochastic simulation of disaster risks in a specified region and examines the ability of the government and private sector to finance relief and recovery. The model is interactive in the sense that the user can change parameters and test different assumptions about hazards, exposure, vulnerability, general economic conditions and the government's ability to respond. As a capacity building tool it can illustrate the tradeoffs and choices government authorities are confronted with for increasing their economic resilience to the impacts of catastrophic events. The model can be used for supporting policy planning processes for the allocation of resources between ex-ante spending on disaster risk management (such as prevention, national reserve funds, sovereign insurance) and ex-post spending on relief and reconstruction. Our paper describes key model features and mechanics, and sets the stage for model applications to the Nepal and Hungary/Tisza cases discussed in this volume.

Keywords Catastrophe modeling • Economic impacts • Government risk management • Fiscal stability • Development

S. Hochrainer-Stigler • G. Pflug
Risk Policy and Vulnerability Department, International Institute for Applied Systems Analysis (IIASA), Schlossplatz 1, A-2361 Laxenburg, Austria

R. Mechler (✉)
Risk, Policy and Vulnerability (RPV) Program, International Institute for Applied Systems Analysis (IIASA), Schlossplatz 1, A-2361 Laxenburg, Austria
e-mail: mechler@iiasa.ac.at

A. Amendola et al. (eds.), *Integrated Catastrophe Risk Modeling: Supporting Policy Processes*, Advances in Natural and Technological Hazards Research 32, DOI 10.1007/978-94-007-2226-2_8, © Springer Science+Business Media Dordrecht 2013

8.1 Introduction

The number of natural disasters and associated losses has been increasing due to population growth and migratory trends from rural to urban areas as well as increases in the value of exposed assets (IPCC 2011; Munich Re 2009). Although climate change is often ascribed to increase the frequency and severity of extreme events, such evidence still remains limited (Solomon et al. 2007). While more developed countries often are well equipped to cope with the impacts of disasters, in less developed countries a much larger proportion of the population is severely affected in terms of loss of life and physical impairment and a substantial strain is put on a country's resources, which may lead to important limitations in the ability to continue financing important social and economic programs (Linnerooth-Bayer et al. 2005).

Historically, losses in developing countries have been funded by diversions of funds from the national budget, loans and donations by the international community. Yet these sources are often insufficient, and ex-post gaps in necessary financing of disaster losses are frequently encountered. As one example, the earthquake of 2001 in the state of Gujarat, India led to a significant shortfall between planned government expenditure, planned funding and actual funding made available (Fig. 8.1).

When stimulus is most needed, such lack of timely funding can lead to important follow-on effects. Observed empirical effects on macroeconomic variables can be summarized as follows (see Mechler 2004; Hochrainer 2006, 2009):

- Compared to more developed economies, significant longer-lasting disaster impacts may be expected depending on the size of event, economic vulnerability, and prevailing economic and socio-political conditions.
- In developing countries, GDP falls in the year of the event or the year after, but rebounds in successive years due to increased investment and capital inflows.
- The public deficit increases due to increased spending needs and decreased tax revenue.
- The trade balance worsens, as imports rise (need for additional goods) and exports fall (destruction of goods produced and productive capital stock) post-catastrophe.
- The inflow of external aid and capital is decisive for the speed of economic recovery.

Our analysis focuses on some of these issues and discusses the need for proper ex-ante planning using catastrophe risk modeling as an important element of a comprehensive disaster risk management approach (Gurenko 2004; World Bank 2008). The discussion presents the IIASA CATSIM (CATastrophe SIMulation) model, which is a model framework to assess country-wide contingent disaster obligations and potential financing shortfalls as well as the costs and benefits of vulnerability – and risk-reduction options. The first version of the model was originally developed in 2002 to inform the *Regional Policy Dialogue* of the Inter-American Development Bank, where it was applied to a number of case

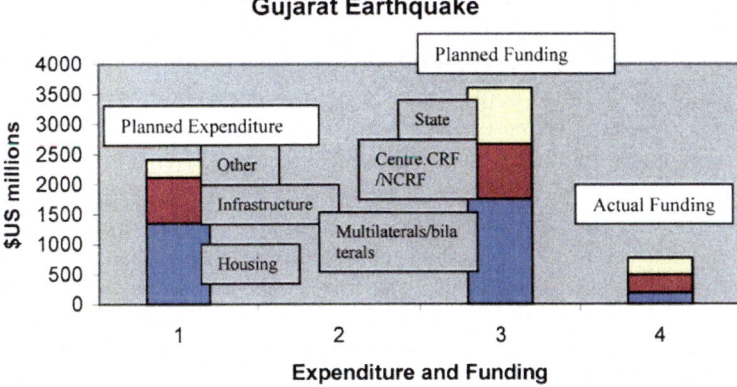

Fig. 8.1 Resource gap in India after Gujarat earthquake (Source: World Bank 2003)

studies in Latin America (see Mechler et al. 2002 Freeman et al. 2002a; Mechler 2004; Hochrainer et al. 2004). The model has since been revised, extended (including the development of a stand-alone application) and utilized by a number of hazard-exposed countries in other regions such as Asia and Africa (see Hochrainer 2006; Mechler et al. 2006; Hochrainer and Mechler 2009).

The discussion on CATSIM in this chapter is organized as follows: Sect. 8.2 discusses the rationale for financing disaster risk. Section 8.3 describes the CATSIM approach and its modeling steps, and Sect. 8.4 ends with conclusions and an outlook to the future. Applications of the model are further discussed in this book in Chap. 9 for the case of Nepal and Chap. 16 for Hungary and the Tisza region.

8.2 The Rationale for Financing Disaster Risk

The rationale for financing disaster risk results from the need of highly exposed countries to protect themselves against resource gaps in dealing with disaster consequences and their associated long-term negative effects. In order to analyze it, one needs first to discuss risk, vulnerability and the exposure of the public sector to disaster risk

8.2.1 Defining Risk and Vulnerability

Risk and *vulnerability* are concepts with multiple and ambiguous meanings. As an analytical term, vulnerability has been confusingly used in an array of disciplinary contexts, including geography, risk and hazard, anthropology, engineering and ecology. Vulnerability is commonly defined in the context of climate change (e.g. IPCC 2007) as a function of both potential impacts and society's capacity to adapt to these impacts. A narrower definition that focuses only on the impacted system is common in the risk/hazards and vulnerability communities. Turner et al. (2003) define vulnerability as the degree to which a system or subsystem is likely to experience harm due to exposure to a hazard, either as a perturbation or a stressor. In this framework, multiple hazards can be caused or aggravated by global-change phenomena, and risk is a function of the hazard (likelihood and severity) and its potential consequences (exposure, vulnerability), but usually fails to consider the coping capacity and resilience (i.e. the ability to return to pre-disaster conditions) of the exposed system. Risk, vulnerability and resilience are important concepts for the model-based analysis of the economic impacts of disasters within the CATSIM model. In the following, we will focus on the concepts of financial and economic vulnerability as well as risk.

8.2.2   Vulnerability and Risk Related to Natural Hazards

The standard approach in catastrophe modeling is to understand natural disaster risk as a function of the hazard, the exposure and the physical vulnerability. Hazard analysis involves determining the type of hazards affecting a certain area with specific intensity and recurrence. Assessing exposure is concerned with analyzing the relevant elements (population, assets) exposed to relevant hazards in a given area. Vulnerability is a multidimensional concept encompassing a large number of factors that can be grouped into physical, economic, social and environmental factors. The factors affecting and comprising vulnerability can be listed as follows (see GTZ 2004).

- Physical vulnerability: factors relate to the susceptibility to damage of engineering structures such as houses, dams or roads. Factors such as demographic change and population growth may also be subsumed under this category.
- Social vulnerability: this can be defined by the ability to cope with impacts on the individual level as well as referring to the existence and robustness of institutions to deal with and respond to natural disaster.
- Environmental vulnerability: a function of factors such as land and water use, biodiversity and stability of ecosystems.
- Economic vulnerability: determinants relate to economic or financial capacity to refinance losses and recover quickly to a previously planned economic activity path. This may relate to private individuals as well as companies and their savings and asset base, or to governments that often bear a large share of country risk and associated losses.

Combining hazard, exposure and physical vulnerability leads to an estimate of *direct* risk in terms of potential effects and losses to be expected. As explained further below, linking direct risk in terms of losses with economic vulnerability produces indirect risk in terms of macro – or microeconomic risk. Risk

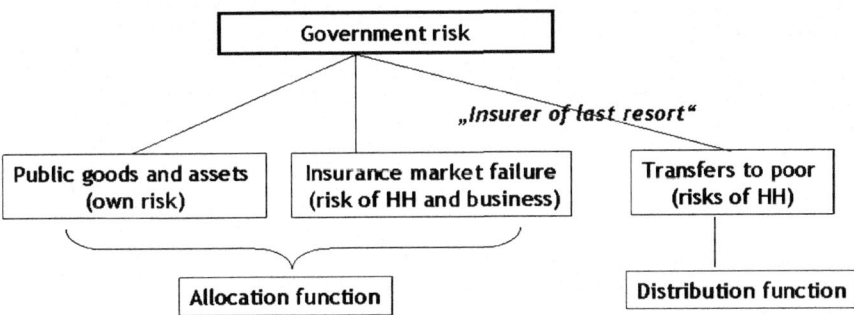

Fig. 8.2 Government disaster risk

management aims at reducing direct and indirect risks. Benefits of risk management are the reduction in risk estimated by comparing the situation with and without risk management. We denote resilience as the ability to return to pre-disaster conditions; appropriate organizational structures, know-how of prevention, risk reduction and response have a decisive influence on resilience.

8.2.3 Fiscal and Economic Implications of Disasters

From an economic perspective, governments are exposed to natural disaster risk and potential losses due to their two main functions: the allocation of goods and services (security, education, environmental protection) and the distribution of income, Schick et al. (2004), as shown in Fig. 8.2.

According to Schick et al. (2004), Stern (2007), in many cases market forces are unlikely to generate adequate adaptation to disaster risks, broadly because of the following three reasons: (1) uncertainty and imperfect information, (2) missing and misaligned markets and (3) financial constraints. In case of a disaster event, consequently, there may be substantial contingent liabilities as identified in Table 8.1. Should governments insure or purchase alternative risk financing instruments for those liabilities? According to an early theorem by Arrow and Lind (1970) a government may

- pool risks as it possesses a large number of independent assets and infrastructure so that aggregate risk is negligible, and/or
- spread risk over the population base, so that per-capita risk is negligible to risk-averse households.

Accordingly governments should behave risk-neutrally and evaluate their investments only through the expected net present (social) value. In theory, thus, governments are not advised to incur the extra costs of transferring their disaster risks if they carry a large portfolio of independent assets and/or they can spread the losses of the disaster over a large population. Because of their ability to spread and

Table 8.1 Government liabilities and disaster risk

	Direct	Contingent
Liabilities	Obligation in *any* event	Obligation if a *particular* event occurs
Explicit Government liability recognized by law or contract	Foreign and domestic sovereign borrowing, expenditureby budget law	State guarantees for non-sovereign borrowing and public and private sector entities, reconstruction of public infrastructure and assets
Implicit A moral obligation of the government	Future recurrent costs of public investment projects, pension and health care expenditure	Default of subnational government and public or private entities, disaster relief to affected households and business

Source: Schick and Polackova Brixi (2004)

diversify risks, Priest (1996) refers to governments as "the most effective insurance instrument of society." Furthermore, the extra costs of insurance can be significant; for example Froot (2001) reports insurance costs of up to seven times greater than the expected loss due to high transaction costs, uncertainties inherent in risk assessment, the limited size of risk transfer markets and the large volatility of losses. According to Arrow and Lind (1970) governments should thus not insure if they are not averse to risks, i.e. if financial risks faced by a government can be absorbed without major difficulty.

The Arrow and Lind theorem has served as the basis for government strategies for dealing with risk. In practice, most governments neglect catastrophic risks in decision making, thus implicitly or explicitly they behave risk-neutrally (Carpenter et al. 2000). The case against risk aversion, however, may not hold for extreme events. As early as 1991, the Organization of American States' primer on natural disasters stated that the risk neutral proposition is valid only up to certain point and that the reality in developing countries suggests that those governments cannot afford to be risk-neutral:

> The reality of developing countries suggests otherwise. Government decisions should be based on the opportunity costs to society of the resources invested in the project and on the loss of economic assets, functions and products. In view of the responsibility vested in the public sector for the administration of scarce resources, and considering issues such as fiscal debt, trade balances, income distribution, and a wide range of other economic and social, and political concerns, governments should not act risk-neutral (OAS 1991).

In these cases governments should justifiably act as risk-averse agents. This means that the Arrow-Lind theorem may not apply to governments of countries that exhibit some or all of the following characteristics (see Mechler 2004):

- high natural hazard exposure;
- economic activity clustered in a limited number of areas with key public infrastructure exposed to natural hazards (see also Hochrainer and Pflug 2009); and
- constraints on tax revenue and domestic savings, shallow financial markets, and high indebtedness with little access to external finance.

These conditions are fundamental for assessing the financial vulnerability of a state. Governments are financially vulnerable to disasters if they cannot access sufficient funding after a disaster to cover their liabilities with regard to reconstructing public infrastructure and providing assistance to households and businesses (Mechler 2004). As an indicator of financial vulnerability, a *resource gap* measures sovereign financial vulnerability *in terms of the lack of sufficient savings or funding for relief and reconstruction.* The repercussions of large resource gaps can be substantial. An inability of a government to repair infrastructure in a timely manner and provide adequate support to low-income households can result in adverse long-term socio-economic impacts. As a case in point, Honduras experienced extreme difficulties in repairing public infrastructure and assisting the recovery of the private sector following Hurricane Mitch in 1998. Five years after Mitch's devastation the GDP of Honduras was 6% below pre-disaster projections.

In considering whether Honduras and other highly exposed countries should protect themselves against resource gaps and associated long-term negative consequences, it is important to keep in mind that risk management measures have associated opportunity costs, which means that they can reduce GDP by diverting financial resources from other public sector objectives, such as undertaking social or infrastructure investments. There are a number of countries like Honduras. Figure 8.3 shows key countries that may need to take a risk averse approach to disaster risk. For this global set of large observed disaster events, losses measured in terms of gross national product are significant for a number of smaller or lower income states, while this ratio becomes smaller for larger and higher income countries.

As one exemplary case, we discuss the case of Nepal in Chap. 9 of this book. Nepal is a country subject to high natural disaster risk and with minor capacity of spreading or pooling the risks. In such circumstances, the Arrow-Lind theorem may not apply, and the argument concerning the risk spreading capacity of governments – and the resulting individual cost being negligible – becomes debatable. In reality, external aid or loans are in dire need post-disaster. In response to evidence and research on the consequences of disasters, a number of developing and transition countries, such as Mexico and Colombia, have modified their reactive approaches to disaster risk and are actively considering risk management and fiscal planning for risk (Cardenas et al. 2007; Linnerooth-Bayer et al. 2011).

8.2.4 Risk Financing Options for Reducing Financial Vulnerability

Governments can choose among a variety of traditional and novel pre-disaster risk financing instruments for reducing their financial vulnerability. The most common are discussed below:

- A *reserve fund* holds liquid capital to be used in the event of a disaster. Ideally, the fund accumulates in years without catastrophes; however, from experience,

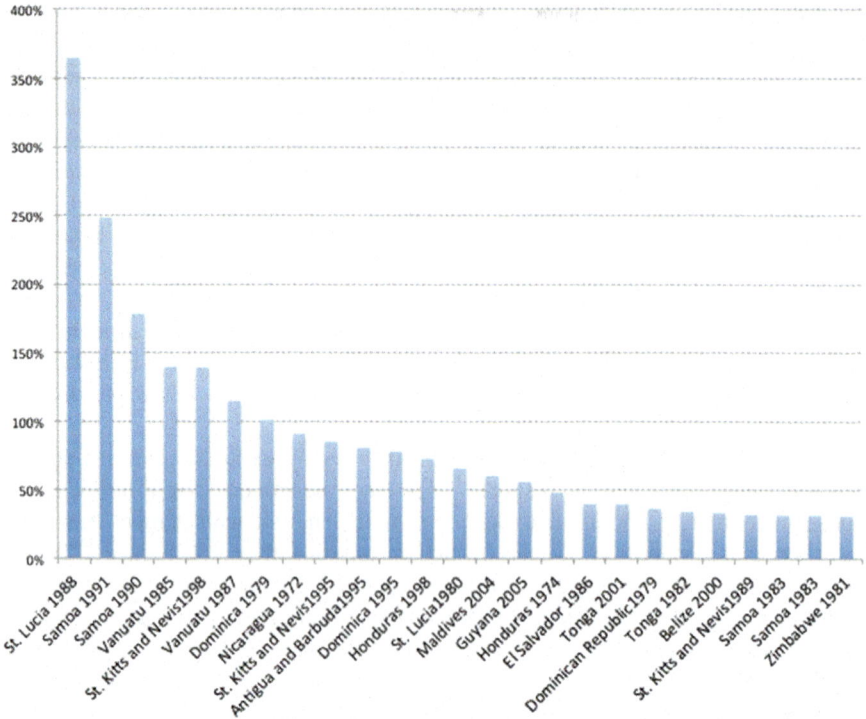

Fig. 8.3 Risk as measured by observed events vs. Gross National Product for large disaster events (Data Source: Mechler et al. 2009; World Bank 2008)

there is considerable political risk of fund diversions to other pressing government needs, especially after long periods without serious disaster impacts.

• *Insurance and other forms of risk transfer* provide indemnification against losses in exchange for a payment. The most common form of risk transfer is insurance or reinsurance. Insurance is an important pre-disaster, risk-transfer institution in that it distributes disaster losses among a pool of at-risk households, businesses and/or governments and to the reinsurance markets. A catastrophe bond (cat bond) is an alternative risk transfer instrument where the investor receives an above-market return when a specific catastrophe does not occur (e.g. an earthquake of magnitude 7.0 or greater), but shares the insurer's or government's losses by sacrificing interest or principal following the event.

• *Contingent credit* arrangements do not transfer risk spatially, but spread it intertemporally. In exchange for an annual fee, the risk cedent has access to a pre-specified post-event loan that is repaid at contractually fixed conditions. In the case of sovereign risk financing, international finance institutions offer such instruments. Contingent credit options are commonly grouped under alternative risk-transfer instruments.

Due to the extreme nature of the losses and the substantial costs involved in such transactions, disaster insurance and other risk financing instruments generally

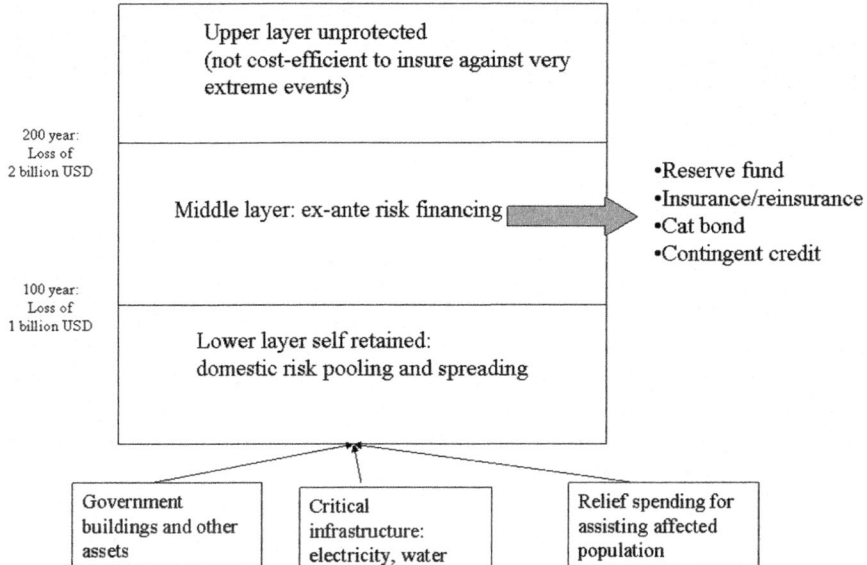

Fig. 8.4 A structure for financially managing the public sector's disaster risk liabilities

absorb only specified layers of risk, defined by attachment and exit points (or lower and upper thresholds based on the recurrence period of the events, as discussed below). Low layers of risk, for which the risk cedent is able to finance the losses, will typically be retained. Extreme layers of risk will also not be transferred to other agents because of the high and exponentially increasing costs of transfer; one important factor is the uncertainty associated with extreme losses: which necessitates large sums of backup capital "reserved" by the agent accepting the risks in order to fulfill her obligation in case of an event.

An example of a layered risk-transfer portfolio is illustrated in Fig. 8.4. In this case, the lower threshold (attachment point) is illustrated as the 100-year event (an event with an annual probability of less or equal to 1%) with losses of $1 billion. The upper threshold (exit point) is the 200-year event with losses of $2 billion. The lower threshold is in principle determined by the government's *financial vulnerability* since it specifies the disaster risk for which the government is in need of additional financial resources for protecting its portfolio of public assets and providing emergency response and relief.

8.3 The CATSIM Model Approach

A number of risk modeling companies are involved in catastrophe risk modeling for insurance and reinsurance companies, to develop adequate financial risk management measures such as estimating required reserve capital or the uptake of reinsurance contracts (Kuzak et al. 2004). In a similar vein, the CATSIM model focuses on the portfolios of governments and outlines the costs and benefits of undertaking risk management options.

8.3.1 Methodology and Structure

CATSIM uses stochastic simulation of a disaster in a specified region and examines the ability of governments and private sectors to finance relief and recovery. It is interactive in the sense that the user can change the parameters and test different assumptions about the hazards, exposure, vulnerability, general economic conditions and the government ability to respond. As a capacity building tool, it can be used to illustrate to the authorities the trade-offs and choices they are confronted with to increase resilience to the risks of catastrophic disasters.

From a methodological perspective, CATSIM approaches the decision and modeling problem as a two stage decision problem under uncertainty. Figure 8.5 outlines the logic followed in the modeling approach. The objective is to guarantee the sufficient and timely financing of government post disaster obligations, the provision of relief to the private sector and the reconstruction of public assets.

In the first, ex ante stage, a part of the government's budget can be allocated to undertake risk reduction (e.g. building a dike), or buy insurance and other financial protection instruments for public assets (such as infrastructure and public buildings) and relief obligations to the private sector. This reduces the budget available for investment into regular development-enhancing activities, creating opportunity costs. The second stage, the decision stage after a disaster, is the ex-post stage where budget reallocation and other financial decisions are made in order to finance the funding needs. Yet, financing the losses with ex-post sources also reduces the budget for investment.

The part of the losses that neither ex-ante nor ex-post options can cover is called *resource gap*. This gap in terms of a shortfall of required resources to continue with key socioeconomic priorities affects key macroeconomic outcomes in the future such as GDP, government revenue and the budget position, and therefore it also increases financial vulnerability and consequently future risks.

8.3.2 Methodological Steps of CATSIM

CATSIM is operationalized in five major steps as described below and illustrated in Fig. 8.6.

Step 1: Risk of direct asset losses (in terms of probability of occurrence and destruction in monetary terms) is modeled as a function of hazards (frequency and intensity), the elements exposed and their physical vulnerability.

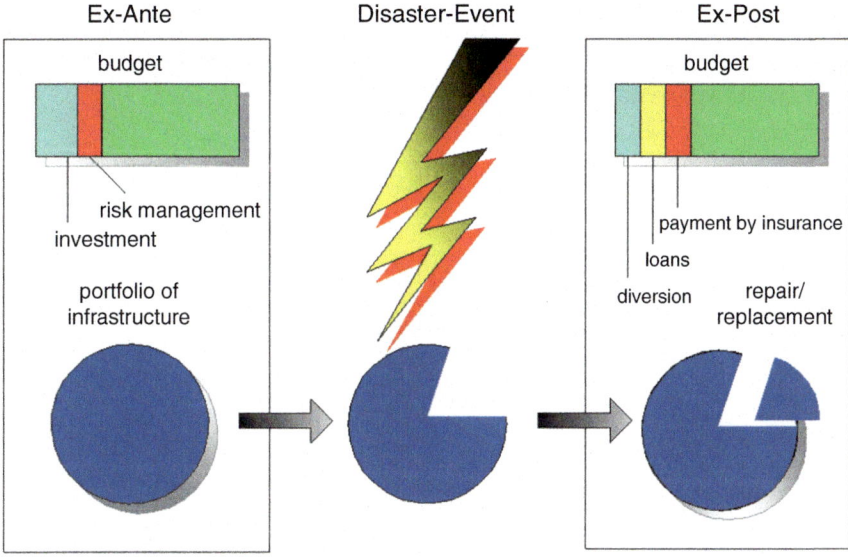

Fig. 8.5 Modeling logic of CATSIM

Step 2: Financial and economic resilience for generally responding to shocks is measured. Resilience is defined as the state or central government's accessibility to savings for financing reconstruction of public infrastructure and providing relief to households and the private sector. Resilience depends heavily on the general prevalent economic conditions of the given country.

Step 3: Financial vulnerability, measured in terms of the potential resource gap, is assessed by simulating the risks to the public sector and the financial resilience of the government to cover its post-disaster liabilities following disasters of different magnitudes.

Step 4: The consequences of a resource gap on key macro variables such as economic growth or the external debt situation are identified. These indicators represent consequences to economic *flows* as compared to consequences to *stocks* addressed by the asset risk estimation in step 1.

Step 5: Strategies can be developed and illustrated that build resilience of the public sector or contribute to the risk management portfolio. The development of risk management strategies has to be understood as an adaptive process where measures are continuously revised after their impact on reducing financial vulnerability and risk has been assessed within the modeling framework.

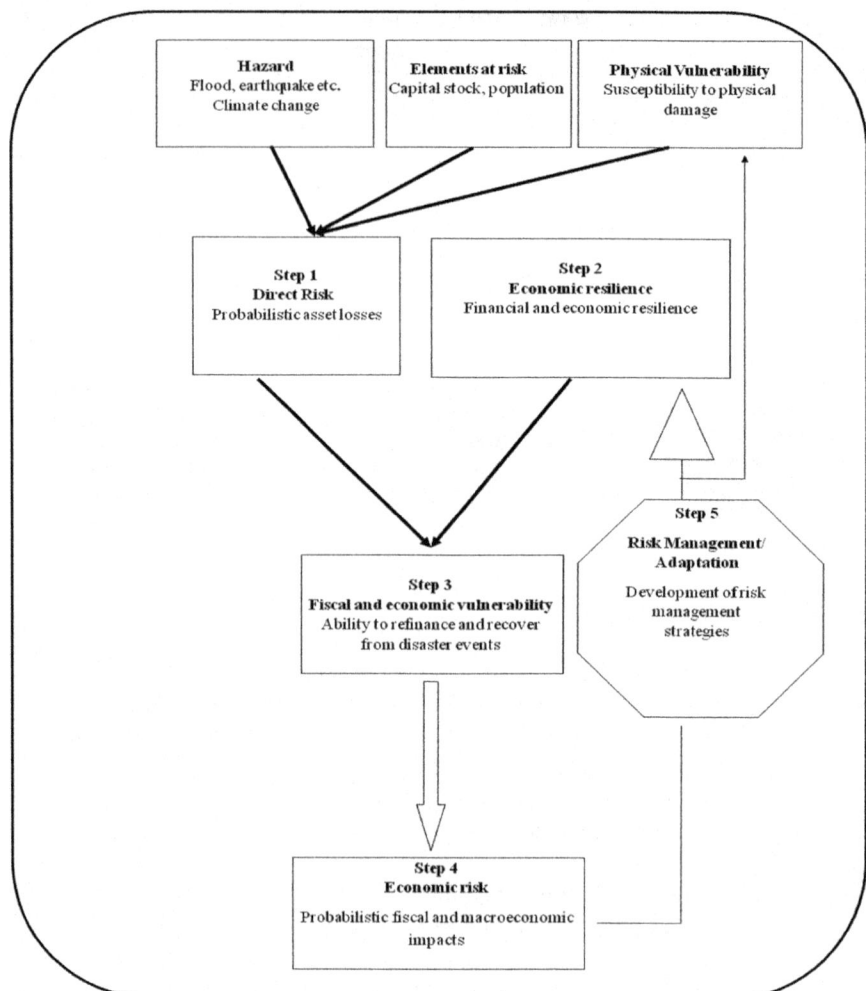

Fig. 8.6 Flow chart of CATSIM methodology

In the next few paragraphs, we discuss each step in more detail.

Step 1: Assessing disaster risk

In the first CATSIM step, the risk of direct losses is assessed in terms of the probability of asset losses in the relevant country or region. Consistent with general practice, risk is modeled as a function of hazard (frequency and intensity), the elements exposed to those hazards and their physical vulnerability (Burby 1991; Swiss Re 2000). In more detail,

- Natural hazards, such as earthquakes, hurricanes, or floods, are described by their intensity (e.g. peak flows for floods) and recurrence (such as a 1 in 100 year events, i.e. with a probability of 1%).
- Exposure of elements at risk is estimated as total private and public capital stock.

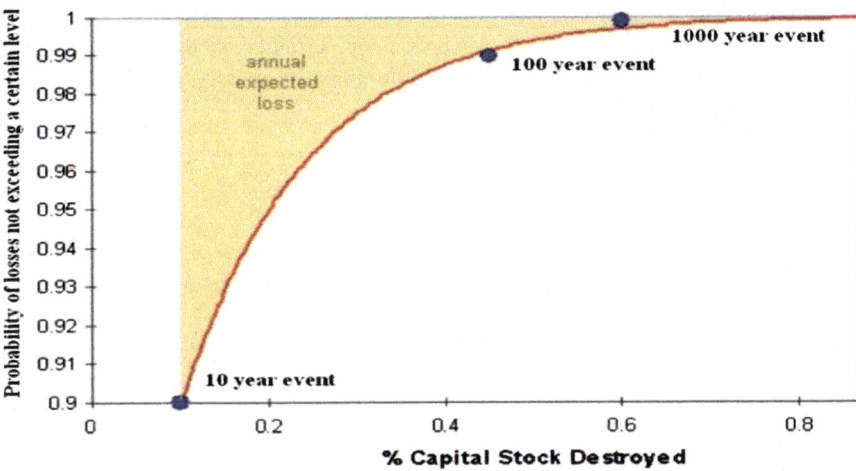

Fig. 8.7 Risk of losses as measured by a cumulative loss-frequency distribution (Source: Freeman et al. 2002b)

- Physical vulnerability describes the degree of damage to the capital stock due to a natural hazard event. The standard method followed here consists of estimating vulnerability or fragility curves putting the degree of losses in relation to the intensity of a hazard.

Based on such information, potential losses due to destructive events can be established for a country, state or region in terms of per cent of capital stock lost. The data on return periods and losses serve as input to CATSIM for generating loss-frequency distributions, which relate probabilities to assets destroyed. For example, Fig. 8.7 shows a cumulative loss-frequency distribution for flood risk in a hypothetical country. The horizontal axis shows the fraction of capital stock destroyed by a disaster, and the vertical axis represents the probability that losses will *not* exceed a given level of damage. For example, with a probability of 0.9 (90%) flood losses will not exceed 0.1% of capital stock; inversely, there is a 0.1 (10% chance) that such a loss and larger will occur.

Top-down estimates at the aggregate national scale are necessarily rough. Since most disasters are rare events, there is usually insufficient historical data at hand; furthermore, it is difficult to include dynamic changes in the system, e.g. change in exposure and hazards due for instance to population and capital movements and climate change. To improve the robustness of estimates bottom-up assessments can be undertaken that involve a detailed analysis of the occurrence of hazards in certain areas, the exposed elements and vulnerabilities of assets at a more detailed scale.

An important summary measure of this distribution is the annual expected loss, the loss to be expected on average every year. The annual expected loss is the sum of all loss weighted by their probability of occurrence. Graphically, the expected losses is represented by the area above the cumulative distribution curve. While the expected loss is an important metric, it should generally be stressed that risk management strategies for extreme events focus strongly on the fat tails of the distribution (the 100 or 200 year events) rather than the average risks.

Step 2: Assessing public sector financial resilience
Based on the information on direct risks to a government portfolio, financial resilience can be evaluated by assessing government's ability to finance its obligations for the specified disaster scenarios. Financial resilience is directly affected by the general conditions prevailing in an economy, i.e. changes in tax revenue have important implications on a country's financial capacity to deal with disaster losses.

The specific question underlying CATSIM is whether a government is financially prepared to repair damaged infrastructure and provide adequate relief and support to the private sector for the estimated damages. For this assessment, it is necessary to examine government's resources including those that can be relied on after the disaster (probably in an ad hoc manner, *ex post* sources) and resources put in place before the disaster (*ex ante* sources). These sources are described below.

Ex post financing sources
The government can raise funds *after* a disaster by accessing international assistance, diverting funds from other budget items, imposing or raising taxes, taking a credit from the Central Bank (which either prints money or depletes its foreign currency reserves), borrowing by issuing domestic bonds, borrowing from international financial institutions and issuing bonds on the international market (Benson 1997a, b, c; Fischer and Easterly 1990). Each of these financing sources can be characterized by costs to the government as well as factors that constrain its availability (Table 8.2). As an example, disaster taxes are not only expensive to administer but may add to recessionary tendencies after large scale disasters (e.g. due to a decrease in consumption).

As a second example, borrowing can also be constrained by existing country debt. CATSIM assumes that the sum of all loans cannot exceed the so-called *credit buffer* for the country. In the *Highly Indebted Poor Countries Initiative* (HIPC) the credit buffer is defined as 150% of the typical export value of this country minus the present value of existing loans (HIPC 2002). These ex post instruments have (often high) associated costs; even budgetary diversions lead to opportunity costs in terms of foregoing other government investments like building health clinics, highways or schools.

Ex ante financing sources
In addition to accessing ex post sources, a government can arrange for financing before a disaster occurs. Ex ante financing options include the instruments discussed above such as reserve funds, traditional insurance instruments (public or private), alternative insurance instruments, or arranging a contingent credit. These ex-ante options can involve substantial annual payments and opportunity costs; statistically the purchasing government will pay more than the expected losses with a hedging instrument than if it absorbs the loss directly. However,

Table 8.2 Ex post financing sources for relief and reconstruction

Type	Source
Decreasing government expenditures	Diversion from budget
Raising government revenues	Taxation
Deficit financing	Central Bank credit
Domestic	Foreign reserves
	Domestic bonds and credit
Deficit financing	Multilateral borrowing
External	International borrowing
	International Aid

under the assumption of risk aversion these measures may still be beneficial depending on the size of potential losses and the degree of financial vulnerability and risk aversion.

Step 3: Measuring financial vulnerability by the **resource gap**

Using the information on direct risks to the government portfolio and financial resilience, financial vulnerability can be evaluated. Financial vulnerability is thus defined as the lack of access of a government to domestic and foreign savings for financing reconstruction investment and relief post-disaster. The shortfall in financing is measured by the term *resource gap*. The term resource gap has been defined in the economic growth modeling literature as the difference between required investments and the actual available resources in an economy.

The main policy recommendation consequently has been to fill this gap with foreign aid (Easterly 1999).[1] Here, this tradition is followed and the resource gap is understood as the lack of financial resources necessary to restore lost assets and continue with development as planned. Figure 8.8 illustrates the calculation of this metric for a hypothetical case.

Given losses due to a certain event, such as the 100 year event (in the example associated with a public sector loss of 4 billion USD), the algorithm evaluates the sources for funding these losses. An implicit ordering of these sources is assumed according to the availability and marginal opportunity costs of the sources: grants would have the least costs associated as these are donations free of cost to the recipient, and thus they would be used first. Second, diversions from the budget could be used, then domestic credit, followed by borrowing from the international institutions (such as World Bank) and the international markets (bonds).

While in this illustration, a 100-year event could be financed, for a 200 year event (public sector loss of 10 billion USD), there would be lack of (ex-post) sources and consequently a resource gap occurs. It is the main objective of CATSIM to illustrate the costs and benefits of closing this government resource gap by ex-ante measures and the consequences of not being able to do so. World

---

[1] This approach has been criticized among others by Easterly (1999) as generally it lacks considering the role of incentives and institutions in economic growth. Nevertheless, it is without doubt that capital investment plays an important role in economic growth.

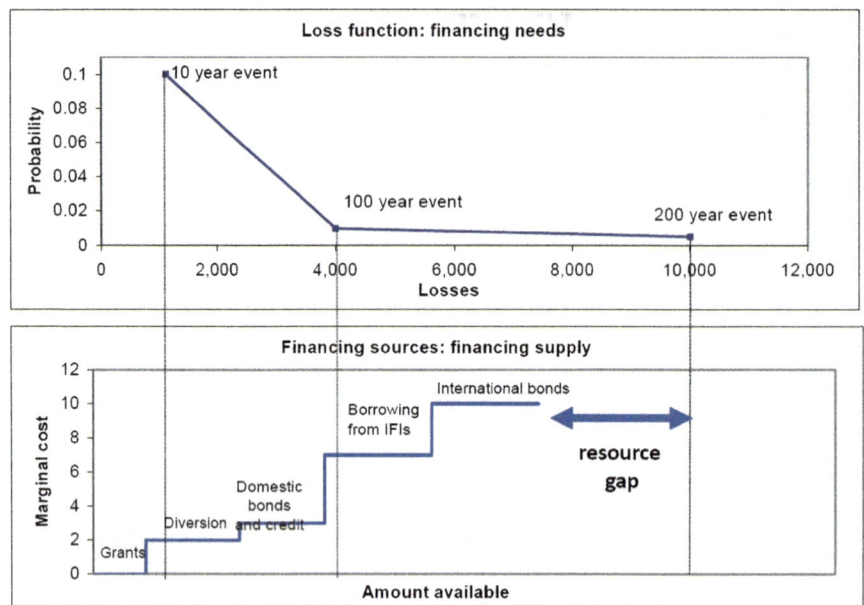

Fig. 8.8 Illustration of methodology for calculating the disaster resource gap (Million USD)

Bank (2008) added another important dimension to this approach in terms of the timing of resource flows. As illustratively shown on Fig. 8.9, while enough funding may become available over time, there may be a temporary resource gap in the aftermath of a disaster event (here shown to be the first 4 months post event) when urgent expenditure needs are high, but immediately available financial resources are often very limited.

While CATSIM is resolved in annual time steps, it considers the fact that the timing of financial inflows for financing the losses is also important and can differ for different ex-ante and ex-post instruments.

Step 4: Illustrating the developmental consequences of a resource gap
Financial vulnerability can have serious repercussions on the national or regional economy and the population. If a government can neither replace nor repair damaged infrastructure (for example, roads and hospitals) nor provide assistance to those in need after a disaster, this will have long-term consequences which can be illustrated by CATSIM. Key aggregate flow outcomes measured by the model are on the fiscal position of a government and the ensuing GDP effects resulting from the lack of ability of a government as a key economic agent to act post event. Governments may brace against these adverse outcomes by implementing physical and financial risk management measures, and generally a government' position and the economy are stabilized against disasters if such measures are adopted. Yet, there are important opportunity costs associated with spending on risk management and in the absence of disaster events, economic welfare will be higher if a government

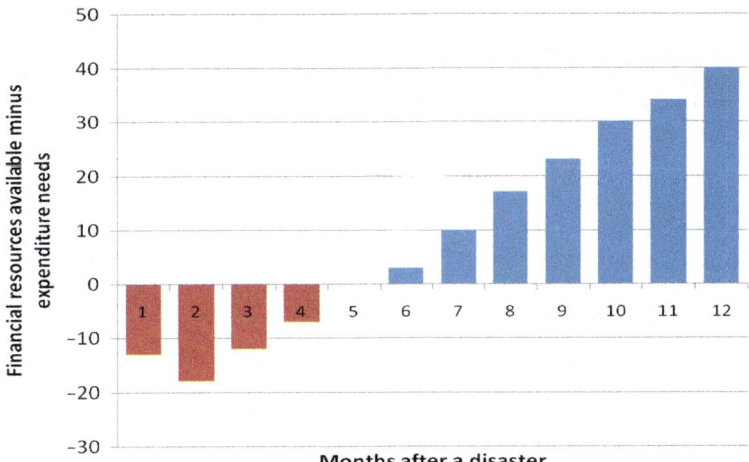

Fig. 8.9 Government resource gap after a natural disaster (Source: World Bank 2008)

does not allocate resources to catastrophe insurance or other risk management measures. This effect is pronounced for financial measures, such as insurance, where annual premiums have to be paid, whereas for physical measures, such as building dikes, the key costs are investment costs to be paid once only.

Step 5: Reducing financial vulnerability and building resilience using ex ante risk management options
Vulnerability and resilience must be understood as dynamic. In contrast to ecological systems, social systems can learn, manage and actively influence their present status quo. There are two types of policy interventions for reducing the financial vulnerability of the public sector: those that reduce disaster risks by reducing exposure and physical vulnerability, and those that build financial resilience of the responding agents. Based on an assessment of the resource gap and potential economic consequences, CATSIM illustrates the pros and cons of strategies for building financial resilience using ex-ante financial instruments. In addition to ex ante financing policy measures (sovereign insurance, contingent credit and reserve funds) one generic option for loss prevention measures has been implemented in order to analyze their linkages with risk financing. Normally, few financial ex-ante options are in place in developing countries, thus the model focuses on analyzing the pros and cons of such new funding sources, which are considered the decision variables.

There are important distinctions between risk reduction and risk financing instruments. While risk financing measures reduce the follow on consequences by transferring risk or sharing risk with others, risk reduction is directed towards decreasing physical vulnerability (Fig. 8.10).

In CATSIM, risk reduction is modeled as an accumulating stock, e.g. similar to a dike used for preventing flooding (see Fig. 8.11). In this representation, there is no

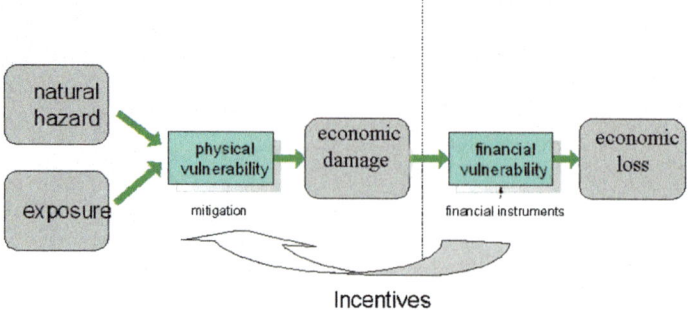

Fig. 8.10 Reducing and financing risk

Fig. 8.11 Model
representation of risk
reduction (Source:
Hochrainer 2006)

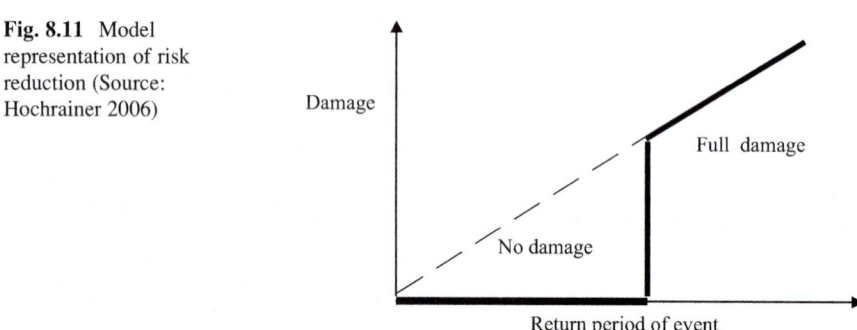

damage if the accumulated risk reduction is able to withstand the theoretical damages due to an event with a certain return period. In Fig. 8.11 the thick line shows the damage as a function of the "hypothetical" damage without risk reduction. No damage occurs up to a given event magnitude. If the magnitude is larger than this limit, the full "hypothetical" damage occurs.

Furthermore, risk reduction and risk financing options are linked within the model. If risk reduction is in effect, it reduces the costs of insurance and contingent credit payments. Thus, risk reduction has the double effect of reducing damage as well as reducing insurance and contingent credit premiums.

8.3.3 Algorithm for Calculating Financing Available from Ex-Ante and Ex-Post Sources

The physical damage is translated into a financial loss for the government after subtracting all ex ante and ex post sources. The existing options are used to the necessary extent. If all of the physical damage can be covered by ex-ante and ex-post options the financial loss is zero. Otherwise, if after exhausting all ex-ante and

Table 8.3 Calculation of ex-ante and ex-post sources

Type of source	Method
Ex-ante	
Insurance	Claim defined by attachment and exit point
Reserve fund	Reserve fund is depleted to the extent necessary up to full depletion
Contingent credit	Triggered to the extent necessary and "reserved in advance" due to payment of a fee for the contingent credit; debt is incurred
Risk reduction	Damages are reduced to zero, if threshold is exceeded full loss occurs and accumulated risk reduction investment is lost
Ex-post	
Budget diversion	Maximum diversion is a fixed percentage of revenue
Aid	Fixed portion of physical loss, assumed to be 10.4% of losses according to statistical analysis done with historical data (see Freeman et al. 2002a)
Domestic credit	Maximum domestic credit available is a fixed fraction of the revenue
Foreign credit	Constrained by external debt sustainability indicator and credit buffer. It is assumed that half of the needed sum comes from multilateral sources and half from issuing international bonds

ex-post sources, there still is a net loss, a resource gap, a part of lost capital stock will remain unreplaced, affecting GDP and leading to lower revenue in the next period. Table 8.3 shows how ex-ante and ex-post instruments resources are determined.

Methodologically, we use lexicographic preference ordering as follows: Let the (monetary) loss distribution for the government be called F. Furthermore, assume that the government has k instruments (either ex-post and/or ex-ante) available to finance the losses. In case of a disaster event some or all of the instruments are used to a given amount to finance the losses. In the simplest case, there is a strict preference order between the financing instruments, represented by the resource vector $\vec{x} = (x_1, \ldots, x_k)'$ in the following way: the first instrument (with monetary resources x_1) is preferred before all others until depletion; afterwards the second instrument (with resources x_2) is preferred before all others until depletion, and so on. Let $\vec{x}_m = (x_{m1}, \ldots, x_{mk})'$ be the maximal (monetary) amount available for each instrument for a given loss event. Then the loss financing scheme for a given event with return period $1/y$ (e.g., for a 100 year event y would be 0.01) is the solution of depleting resources in the respective order till the losses ($F^{-1}(1 - y)$) are fully financed. In case that $\sum_{i=1}^{k} x_{mi} < F^{-1}(1 - y)$ is fulfilled, a resource gap occurs, and the return period of the event where this happens for the first time (i.e., all events with lower return periods satisfying equation $\sum_{i=1}^{k} x_{mi} = F^{-1}(1 - y)$) is called *the critical return period*. As indicated, resource gaps will have (possibly long-term) economic consequences, which are assessed by the economic module discussed next.

8.3.4   The Economic Module

Financial resilience is part and parcel of the general conditions of the modeled economy and is analyzed independently of disaster risk. In CATSIM, the macro-economic module is currently set out as a simple Solow-type growth framework with the focus on the potential for medium to longer term growth and development of aggregate economic variables given explicit consideration of disaster risks (see Barro and Sala-I-Martin 2004 for a discussion of the economic growth literature). The Solow model (more correctly *Solow-Swan model*) is considered the workhorse of economic growth research for studying the longer term potential development of an economy. In the simple exogenous savings version, economic growth is driven by the accumulation of capital via the savings-investment relationship and the rate of depreciation. Modeling economic growth only as a function of capital stock and the availability of new investment into capital stock has to be regarded as a limitation of the model. Solow and others have shown in the 1950s that in advanced countries more than 50% of economic growth can be explained by productivity increases. This number may not be as large for developing countries, but suggests that a considerable amount of growth is not purely driven by the amount of capital but rather its quality (Dinwiddy and Teal 1996). Also, today's economic theory generally stresses the importance of incentives, the role of human and social capital and the importance of robust institutions for economic development (Meier 1995). On the other hand, it is generally acknowledged that capital investment plays a major role as a driver of economic growth. CATSIM makes a number of important modifications to the model:

- The main focus is on the public sector (national or state government), its fiscal liabilities and risk management strategies; the model is solved accordingly.
- Capital can be destroyed by natural disasters. As the occurrence of disasters is modeled stochastically, stocks and flows such as assets, budget and GDP become stochastic variables (labor is currently fixed).
- The private and public sector investment budget can be used for investing in new capital stock (or maintaining existing), replacing destroyed stocks or for protecting these assets by the ex-ante risk reduction measures or risk financing.
- There is a fixed government budget to be used for consumption and investment. Reconstruction of destroyed stocks has to be financed from the budget as well. Also debt service payments (e.g. due to incurring new debt for purposes of reconstruction) have to be paid from this budget.
- The investment budget can be used for investing in new capital stock (or maintaining existing) or for protecting these assets by the ex-ante risk manage-ment measures risk reduction or risk financing.

Table 8.4 gives an overview of important model components as part of the modeling approach.

The purpose of the economic module is not to develop estimates for the main economic variables, but to contrast a baseline to a case with additional ex-ante

Table 8.4 Overview of important model features of CATSIM approach

Model feature	Description
Assumed government objectives	Provide relief post-disaster and rebuild infrastructure quickly while maintaining growth
GDP growth	Endogenous, GDP falls in year of event, in subsequent years GDP is determined by investment in previous year
Reconstruction investment	Government undertakes reconstruction investment for infrastructure, private sector undertakes reconstruction investment for private capital
Domestic savings	Limited supply, decrease after event, as income falls
Government consumption	Constant except for year of catastrophe
Private consumption	Constant, as low per capita income households increase their propensity to consume to maintain life-sustaining level of spending
Production function	Cobb-Douglas with inputs capital and labor
Treatment of capital	Catastrophe destroys capital
Treatment of labor	Labor force decreased in year of event
Imports and exports	Closed economy assumption

protection for disaster risk and study the associated effects over a certain time horizon. We use a production function approach which seems most suitable for this purpose. Currently, in order to represent the production of goods (supply) a simple Cobb-Douglas function is used with inputs capital and labor.

$$GDP = AK^{\alpha}L^{\beta}$$

where K represents capital stock, L effective labor force, A is a technological efficiency parameter, alpha and beta represent the production elasticity of capital stock and labor.

8.3.5 Representing Uncertainty

Another key issue for CATSIM is the analysis of uncertainty (see also Compton et al. 2009 and Chap. 2 in this book). Three types of uncertainties are considered: *aleatoric* uncertainty, *parametric* uncertainty and *model* uncertainty. While model uncertainty (the uncertainty that the model appropriately represents the actual system) is more difficult to tackle and based on modeler's choices (see also Chap. 9 in this book for the case of Nepal), aleatoric uncertainty (natural variability) is considered by the above mentioned loss-frequency distributions. Because of the simulation approach used, response variables are expected values and it is important to determine the parametric uncertainty around these estimates. Confidence intervals are used to reflect this uncertainty.

Another type of uncertainty, *epistemic* uncertainty (for a discussion of uncertainties see Chap. 2 by Compton et al. in this book), is harder to treat mathematically. Usually,

Fig. 8.12 CATSIM in use to inform planning for disasters with officials from Caribbean countries, Barbados, June 2006

the mathematical treatment of *epistemic* uncertainties requires encoding and aggregation of expert opinions. Different approaches for aggregation exist, however, various problems arise due to the issues involved in the weighting process (Pate-Cornell 1996), and thus cannot be seen as very reliable proxies for this kind of uncertainty. In our approach, this kind of uncertainty is dealt with more broadly by involving key stakeholders from finance ministries, disaster management authorities or civil society in deliberative processes organized around workshops. These workshop are facilitated by a standalone software version of CATSIM, which is equipped with a graphical user interface making it possible to systematically assess expert opinions (Fig. 8.12).

This approach allows users to change important parameters and assumptions and study the consequences. Furthermore, as we understand the problem of government risk financing as a trade-off, this setup allows the user to decide which trade-off he/she is willing to commit to and which indicators he/she considers most useful for analyzing the trade-off. The case study on Nepal presented in the Chap. 9 gives some insight into these modeling and decision structuring elements.

8.4 Conclusions

Governments of developing and transition countries frequently face post – disaster resource gaps in financing response, relief, and reconstruction, which can have serious effects on longer-term socioeconomic development prospects. The potential for a resource gap and associated adverse consequences provides a rationale for these countries – overriding the Arrow-Lind theorem – to behave as risk-averse agents and consider risk financing options for their contingent disaster liabilities. Risk financing may be implemented using instruments such as catastrophe reserve funds, sovereign insurance, catastrophe bonds or contingent credit contracts. CATSIM informs this decision problem, and we suggest, its modeling approach focusing specifically on risk as well as the translation of direct to indirect risk can be useful input for informing planning decisions related to sovereign disaster risks. Also, while in this paper different risk management options were assessed separately, more realistic, as well as mixed, strategies can be analyzed with CATSIM, e.g. spending a portion of the budget on risk reduction and insurance. For such an analysis, additional information on the preferences and strategies of decision makers are necessary. To elicit those in interaction with potential stakeholders, the model has been used for a number of applications and workshops sponsored by international organizations involved in disaster risk management, which confirmed the validity of the assumptions and its usefulness for developing sound risk management strategies. Due to the user interface and its philosophy of using simulation rather than optimization analysis, the flexibility to consider multiple aspects in informing decisions constitutes a very important feature of the model. For example, for the case of Mexico, which insured its liabilities in 2006, CATSIM provided information on the different layers of seismic risk to the public finances and helped identify which risks could be transferred to the international markets at an acceptable cost (Cardenas et al. 2007). Yet, finally, the government insured its potential post-disaster relief expenditure based on the fact that congress appropriations for a national reserve trust fund had been volatile and subject to political intervention. Thus, beyond economic efficiency, timing and equity considerations, the key objective of the transaction was to achieve security for the planning process. Clearly, any decision making process will depend on such and many other factors, including expert as well as subjectively constructed information, and we propose to embed CATSIM in deliberative processes involving workshops with both stakeholders and experts leading to mutual learning and hopefully improved decisions.

References

Arrow KJ, Lind RC (1970) Uncertainty and the evaluation of public investment decisions. Am Econ Rev 60:364–378

Barro RJ, Sala-I-Martin X (2004) Economic growth. MIT Press, Cambridge

Benson C (1997a) The economic impact of natural disasters in Fiji. Overseas Development Institute, London

Benson C (1997b) The economic impact of natural disasters in Viet Nam. Overseas Development Institute, London

Benson C (1997c) The economic impact of natural disasters in the Philippines. Overseas Development Institute, London

Burby R (1991) Sharing environmental risks: how to control governments' losses in natural disasters. Westview Press, Boulder/Colorado

Cardenas V, Hochrainer S, Mechler R, Pflug G, Linnerooth-Bayer J (2007) Sovereign financial disaster risk management: the case of Mexico. Environ Hazards 7(1):40–53

Carpenter G in association with EQECAT with contributions from IIASA (2000) Managing the financial impacts of natural disaster losses in Mexico. Mexico Country Office/World Bank, Washington, DC/Latin and Caribbean Region

Compton KL, Faber R, Ermolieva T, Linnerooth-Bayer J, Nachtnebel HP (2009) Uncertainty and disaster risk management; modeling the flash flood risk to Vienna and its subway system. International Institute for Applied Systems Analysis. IIASA Report RR-09-002, Laxenburg, Austria

Dinwiddy C, Teal F (1996) Principles of cost-benefit analysis for developing countries. Cambridge University Press, Cambridge

Easterly W (1999) The ghost of financing gap: testing the growth model used in the international financial institutions. J Dev Econ 60:60:424 ff

Fischer S, Easterly W (1990) The economics of the government budget constraint. World Bank Res Obs 5:127–142

Freeman PK, Martin LA, Linnerooth-Bayer J, Mechler R, Saldana S, Warner K, Pflug G (2002a) Financing reconstruction phase II background study for the inter-American development bank regional policy dialogue on national systems for comprehensive disaster management. Inter-American Development Bank, Washington, DC

Freeman PK, Martin L, Mechler R, Warner K, Hausman P (2002b) Catastrophes and development, integrating natural catastrophes into development planning, Disaster risk management working paper series no.4. World Bank, Washington, DC

Froot KA (2001) The market for catastrophe risk: a clinical examination. J Financ Econ 60:529–571

GTZ (2004) Risk analysis: a basis for disaster risk management guidelines. GTZ, Eschborn

Gurenko E (2004) Catastrophe risk and reinsurance: a country risk management perspective. Risk Books, London

HIPC (2002) About the HIPC initiative, Washington, DC. http://www.worldbank.org/hipc/about/hipcbr/hipcbr.htm

Hochrainer S (2006) Macroeconomic risk management against natural disasters. German University Press, Wiesbaden

Hochrainer S (2009) Assessing macroeconomic impacts of natural disasters: are there any? Policy research working paper, 4968. World Bank, Washington, DC

Hochrainer S, Mechler R (2009) Assessing financial and economic vulnerability to natural hazards: bridging the gap between scientific assessment and the implementation of disaster risk management with the CatSim model. In: Patt A, Schröter D, Klein R, de la Vega-Leinert A (eds) Assessing vulnerability to global environmental change. Earthscan, London

Hochrainer S, Pflug G (2009) Natural disaster risk bearing ability of governments: consequences of kinked utility. J Nat Disaster Sci 31:11–21

Hochrainer S, Mechler R, Pflug G (2004) Financial natural disaster risk management for developing countries. In: Proceedings of XIII, Annual conference of European Association of Environmental and Resource Economics, Budapest

IPCC (2007) Working group II contribution to the intergovernmental panel on climate change fourth assessment report climate change 2007: climate change impacts, adaptation and vulnerability summary for policymakers. IPCC, Geneva

IPCC (2011) Summary for policymakers. In: Field CB, Barros V, Stocker TF, Qin D, Dokken D, Ebi KL, Mastrandrea MD, Mach KJ, Plattner G-K, Allen S, Tignor M, Midgley PM (eds) Intergovernmental panel on climate change special report on managing the risks of extreme events and disasters to advance climate change adaptation. Cambridge University Press, Cambridge/New York

Kuzak D, Campbell K, Khater M (2004) The use of probabilistic earthquake risk models for managing earthquake insurance risks: example for Turkey. In: Gurenko E (ed) Catastrophe risk and reinsurance: a country risk management perspective. Risk Books, London, pp 41–64

Linnerooth-Bayer J, Mechler R, Pflug G (2005) Refocusing disaster aid. Science 309:1044–1046

Linnerooth-Bayer J, Mechler R, Hochrainer-Stigler S (2011) Insurance against losses from natural disasters in developing countries. Evidence, gaps and the way forward. J Integr Disaster Risk Manag 1(1):1–23

Mechler R (2004) Natural disaster risk management and financing disaster losses in developing countries. Verlag fuer Versicherungswissenschaft, Karlsruhe

Mechler R, Pflug G (2002) The IIASA model for evaluating ex-ante risk management: case study Honduras. Report to IDB. IDB, Washington, DC

Mechler R, Linnerooth-Bayer J, Hochrainer S, Pflug G (2006) Assessing financial vulnerability and coping capacity: the IIASA CATSIM model. In: Birkmann J (ed) Measuring vulnerability and coping capacity to hazards of natural origin: concepts and methods. United Nations University Press, Tokyo, pp 380–398

Mechler R, Hochrainer S, Pflug G, Lotsch A, Williges K (2009) Assessing the financial vulnerability to climate-related natural hazards. Background paper for the development and climate change world development report 2010. Policy research working paper 5232. World Bank, Washington, DC

Meier GM (1995) Leading issues in economic development. Oxford University Press, New York

Munich Re (2009) Topics Geo. Natural catastrophes 2008: analyses, assessments, positions. Munich Reinsurance Company, Munich

OAS (1991) Primer on natural hazard management in integrated regional development planning. Organization of America, Washington, DC

Pate-Cornell E (1996) Uncertainty in risk analysis: six levels of treatment. Reliab Eng Syst Saf 54:95–111

Priest GL (1996) The government, the market, and the problem of catastrophic loss. J Risk Uncertain 12:219–237

Schick A, Polackova Brixi H (eds) (2004) Government at risk. World Bank/Oxford University Press, Washington, DC

Solomon S, Qin D, Manning M, Chen Z, Marquis M, Averyt KB, Tignore M, Miller HL (eds) (2007) Climate change 2007: the physical science basis. Contribution of working group I to the fourth assessment report of the intergovernmental panel on climate change. Cambridge University Press, Cambridge

Stern N (2007) The economics of climate change: the stern review. Cambridge University Press, Cambridge

Swiss Re (2000) Storm over Europe: an underestimated risk. Swiss Reinsurance Company, Zurich

Turner BL et al (2003) A framework for vulnerability analysis in sustainability science. PNAS 100:8074–8079

World Bank (2003) Financing rapid onset natural disaster losses in India. A risk management approach. World Bank, Washington, DC

World Bank (2008) Catastrophe risk financing in developing countries. Lessons learned and principles for public intervention. World Bank, Washington, DC

Chapter 9
Managing Indirect Economic Consequences of Disaster Risk: The Case of Nepal

Reinhard Mechler, Stefan Hochrainer-Stigler, and Kazuyoshi Nakano

Abstract Natural disasters can exert significant economic and developmental impacts in countries that lack the economic resilience to bounce back post event. Yet, the brunt of these impacts often goes unrecorded and the information base for improving the financial and economic management of disaster risk in many instances is at best limited. Systematic disaster risk modeling can be a starting point for devising a comprehensive risk management approach. This chapter presents quantitative modeling analysis using the IIASA CATSIM framework for assessing economic natural disaster risk for the case of Nepal. We calculate country level direct disaster risk as well as the corresponding indirect effects using growth modeling and input-output analysis. We find the economic and fiscal risks posed by natural disasters in Nepal to be large and potentially long-lasting, particularly when they are triggered by earthquake risk. As well, disaster events ripple through the economy and may lead to important distributional effects. Given these results, we suggest there is a clear case for considering risk in economic and fiscal planning processes in Nepal and similar heavily disaster exposed countries.

Keywords Fiscal and economic risks • Disaster risk modeling • Multi-risk assessment • Extreme events • Nepal

R. Mechler (✉) • S. Hochrainer-Stigler
Risk, Policy and Vulnerability (RPV) Program, International Institute for Applied Systems Analysis (IIASA), Schlossplatz 1, A-2361 Laxenburg, Austria
e-mail: mechler@iiasa.ac.at

K. Nakano
Disaster Prevention Research Institute, Kyoto University, Kyoto, Japan

A. Amendola et al. (eds.), *Integrated Catastrophe Risk Modeling: Supporting Policy Processes*, Advances in Natural and Technological Hazards Research 32,
DOI 10.1007/978-94-007-2226-2_9, © Springer Science+Business Media Dordrecht 2013

9.1   Assessing the Economic Impacts of Disasters in Nepal

Economic and developmental impacts of disasters in Nepal have been reported to be large and significant (Upreti 2006). Yet, as in many other heavily exposed, developing countries, the brunt of the impacts is hidden, very little quantitative data is available and there has not been systematic economic analysis of the repercussions of disaster risk. To tackle these issues and provide a more robust base for decisions, a consortium composed of the Asian Disaster Preparedness Centre (ADPC), the Norwegian Geological Institute (NGI) and IIASA developed an all-hazards risk model for Nepal to help inform effective risk management strategies (see ADPC, NGI and IIASA 2010). This chapter reports on one project element, the modeling of the indirect, economic risks of natural disasters in terms of potential fiscal and macroeconomic impacts using the IIASA Catastrophe Simulation Model (CATSIM) (for a model description see also Hochrainer-Stigler et al. 2012 in this volume). While estimates of direct risk based on an assessment of hazard, exposure, vulnerability and finally direct risk provide useful information, we argue – and demonstrate this for the case of Nepal – that information on economic risk, combining direct risk estimates with economic vulnerability and resilience, is key for informing decisions. In our analysis, we calculate country level direct disaster risk as well as the corresponding indirect effects using growth modeling and input-output analysis. We find the economic and fiscal risks triggered by natural disasters for Nepal to be large and potentially long-lasting, particularly when they are triggered by earthquake risk. Given these results, we suggest there is a clear case for considering risk in economic and fiscal planning processes in Nepal and similar heavily disaster exposed countries.

The chapter is organized as follows: In Sect. 9.2, we discuss the socio-economic context and the burdens imposed by disasters on Nepal. In Sect. 9.3, we introduce a number of options for better handling disasters, which may be informed by our modeling approach. Before we turn to discussing the application of CATSIM to the given case (Sect. 9.5), we introduce the main modeling steps in Sect. 9.4. We end with a discussion of results and some concluding remarks.

9.2   The Burden of Natural Disasters in Nepal

Natural hazards in Nepal are associated with large direct losses and significant burdens to development. While the direct impacts in theory can be readily observed (in practice they are often not systematically recorded), many indirect effects go unnoticed. These are difficult to observe and generally limited data is reported. In the following we discuss what is available in terms of data regarding socioeconomic characteristics of the country and information on direct and indirect observed impacts.

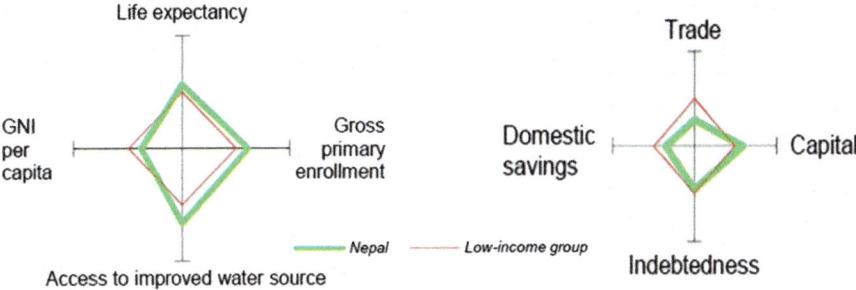

Fig. 9.1 Key development and economic indicators for Nepal as compared to the low-income group (Source: World Bank 2009)

9.2.1   Socioeconomic Context

Nepal is classified as a low income country with a high prevalence of poverty. Population under the poverty line recently exceeded 30%. Income per capita in 2008 was only about USD 250 (in constant 2000 USD) and is low even when compared to the low-income group of countries as defined by the World Bank. Also, domestic savings are low compared to the other countries classified by the World Bank as belonging to the low-income group (Fig. 9.1) (World Bank 2009).

Key socio-economic indicators and data for Nepal for the year 2008 are summarized in Table 9.1. While the World Bank classifies Nepal as a less indebted country, the level of external debt is high compared to the public budget. As one important indicator, the present value of external debt as measured against revenue excluding grants exceeded 247% in 2008. This means that the amount of debt which the government can additionally borrow from abroad is rather limited given that 250% for this indicator is often considered a threshold for debt sustainability (see HIPC 2002). Almost all of the external borrowing in Nepal is done by the central government. Most of the external debt is extended by multilaterals such as the World and Asian Development Banks. According to the World Development Indicators, external debt in the private sector was zero in 2008, and the private sector did not have access to international financial markets.

9.2.2   The Direct and Indirect Burden of Natural Disasters

Nepal is exposed to many different types of natural hazards, including earthquakes, floods, droughts, landslides and epidemics. Impact data, albeit with limited coverage, exists on the direct losses, fatalities and people affected. Looking back over the last 110 years (1901–2010) for which data exists, earthquakes and flooding can be

Table 9.1 Socio-economic indicators for Nepal (year 2008)

Social indicators	
Population	28 million
Surface area	147,180 km^2
Population density/km2	200
Population growth	1.8%
Life expectancy at birth	66
Infant mortality (per 1,000 live births)	40
Poverty (% of population below national poverty line)	31
Economic indicators	
GDP	7.3 billion USD
GDP per capita	253
GDP growth	5.3%
Fiscal revenue excluding grants	12.2% of GDP
Fiscal revenue including grants	15.8% of GDP
Inflation, consumer prices	10.9%
Present value of external debt (current USD)	2.2 billion USD
Present value of external debt	247% of revenue excluding grants

Source: World Bank (2009)
USD values in 2000 USD terms if not indicated otherwise

Table 9.2 Top 10 natural disasters in terms of people killed, affected and losses over the period 1901–2010

Disaster	Date	Killed	Disaster	Date	Affected	Disaster	Date	Losses[a]
Earthquake	1934	9,040	Drought	1979	3,500,000	Flood	1987	728
Epidemic	1991	1,334	Drought	1972	900,000	Earthquake	1980	246
Flood	1993	1,048	Flood	2004	800,015	Flood	1993	200
Epidemic	1950	1,000	Flood	2007	640,706	Earthquake	1988	60
Flood	1996	768	Flood	1993	553,268	Flood	2009	60
Earthquake	1988	709	Flood	1987	351,000	Flood	1998	22
Flood	1981	650	Earthquake	1988	301,016	Drought	1972	10
Epidemic	1992	640	Landslide	2002	265,865	Flood	1983	10
Landslide	2002	472	Earthquake	1980	240,600	Flood	2000	6.3
Flood	1970	350	Flood	1983	200,050	Wildfire	1992	6.2

Source: CRED (2010)
Note: Economic losses in this table means direct asset losses, which is not the same as the indirect, economic impacts discussed further below
[a]In million current USD

said to represent the largest threats to human life and economic assets. We focus on these two hazard types in the following model-based analysis. Table 9.2 summarizes the top ten disasters in terms of people killed, affected and direct losses over that time period.

Disasters are unevenly spread out over Nepal, and the highest impacts in terms of loss of life and loss of assets have been reported for Eastern Nepal, where a large part of the economic assets and the population is located (Upreti 2006).

As indicated above, there is little reported quantitative evidence regarding the full burden imposed by disasters in Nepal (see Upreti 2006), and next to nothing is known

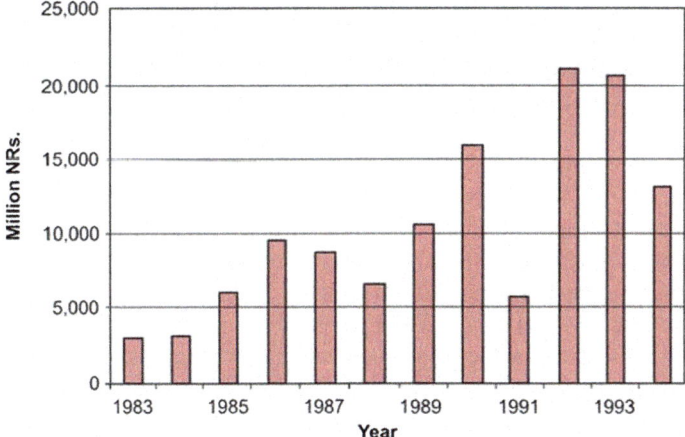

Fig. 9.2 Foreign assistance for disaster relief spending from 1983 to 1994 (Source: Upreti 2006). Note: In million Nepalí rupees. In 1994, 49.4 Nepalí rupees were equivalent to 1 USD (World Bank 2012)

about the economic implications including the macroeconomic impacts. It is known, however, that being a resource-constrained and least developed country (LDC), Nepal has been suffering disaster losses in the range of several billion rupees annually over the last few years. Also, the government of Nepal is already highly dependent on foreign assistance and lending, and about a third of the country's income stems from foreign aid with 64% of development spending disbursed by multilaterals (Bhattarai and Chhetri 2001). It comes as no surprise that foreign assistance for disasters has been large and rising as documented in Fig. 9.2 for the period 1983–1994, for which data was available.

Given a fragile economy exposed to substantial disaster risk, the government of Nepal is faced with developing far-reaching strategies, programmes and projects in order to ensure that key development and poverty alleviation objectives are achieved. It seems important to inform such planning by estimates of natural disaster risk and its potential economic repercussions. A number of entry points for better planning and managing risk exist, and in the following we discuss important disaster policy options, and particularly refer to those which our modeling approach may inform.

9.3   Policy Options for Managing Disaster Risk

9.3.1   Overview

Many options for reducing and managing disaster risk are available, in principle. After the fact (*ex post*) efforts such as providing relief and reconstruction are ever important and still dominate the field. Yet, as resources post-disaster are often very limited and risks are increasing, there is a critical need for stronger implementation of *ex ante* measures, which can be subsumed under the headings of risk assessment, risk prevention, preparedness and risk financing (see Table 9.3). All options need robust input from various sources, and those measures that are shaded in grey can be informed by our model-based analysis presented in this chapter.

In terms of ex ante approaches involving an estimate of risk, impact/risk assessment and planning can be informed using economic risk analysis; importantly, if risk is considered to be large, it should be mainstreamed into development. For risk prevention, economic incentives play a role, which can be informed by estimates of the developmental cost of disasters as well as the benefits of reducing those; education and awareness-raising is another important category for ensuring risk prevention. Risk sharing and financing options need to be based on economic analysis in order to properly identify which risks to keep, finance or transfer. In terms of ex post approaches, often involving an (deterministic) analysis of impacts, response options can be informed by economic loss assessments, consequently leading to the mobilization of financial and other recovery resources from sources such as the public sector, multilaterals or the insurance sector. For matters of reconstruction and rehabilitation, economic modeling can be helpful for designing options for revitalizing affected sectors such as tourism, agriculture, exports etc., as well as for sound macroeconomic and budget management in order to stabilize and protect social expenditures.

9.3.2 Planning Economic Risk

Our analysis focuses to a large extent on risk assessment and planning aspects of risk management. As one entry point, disaster risk can be incorporated into development plans and mainstreamed into strategies, plans, the budget, policies, regulations, programs and finally projects as shown in Fig. 9.3.

This analysis refers to integrating disaster risk with budget planning; the associated planning problem is one of contingency liability planning with fiscal disaster risk resulting from explicit and implicit contingent public sector liabilities. Before explaining in Sect. 9.4 how this problem is operationalized in CATSIM, we shortly discuss the relevance of risk for this as well as other problems in the context of decision making under risk.

9.3.3 Relevance of Risk for Assessing Options

Determining how much should be invested in risk reduction and how much in risk financing is not a straightforward proposition. It ultimately depends on the wider costs and benefits of both types of activities, on their interaction (e.g., financial instruments, through incentives, can influence prevention activities, see

Table 9.3 Overview of disaster risk management measures

Type	Ex ante risk management				Ex post disaster management	
	Risk assessment and planning	Prevention	Preparedness	Risk sharing and financing	Response	Reconstruction and rehabilitating
Effect	Assessing risk	Reducing risk by addressing underlying factors	Reducing risk in the onset of an event	Transferring risk by reducing variability and longer term consequences	Responding to an event	Rebuilding and rehabilitating post event
Key options	Hazard assessment and monitoring	Physical and structural risk reduction works (e.g. irrigation, embankments)	Early warning systems, communication systems	Risk transfer (by means of (re-) insurance) for public infra-structure and private assets, microinsurance	Humanitarian assistance	Rehabilitation/ reconstruction of damaged critical infrastructure
	Vulnerability assessment (population and assets exposed)	Land-use planning and building codes	Emergency response	Alternative risk transfer	Clean-up, temporary repairs and restoration of services	Revitalization for affected sectors (tourism, agriculture, exports etc.)
	Risk assessment as a function of hazard, exposure and vulnerability	Economic incentives for proactive risk management	Networks of emergency responders	National and local reserve funds	Loss assessments	Macroeconomic and budget management (stabilization, protection of social expenditures)
	Mainstreaming risk into development planning	Education, training and awareness raising about risks and prevention	Shelter facilities and evacuation plans	Calamity Funds (national or local level)	Mobilization of recovery resources (public/ multilateral/ insurance)	Incorporation of disaster mitigation components in reconstruction activities

Note: options marked in grey can be informed by the present analysis

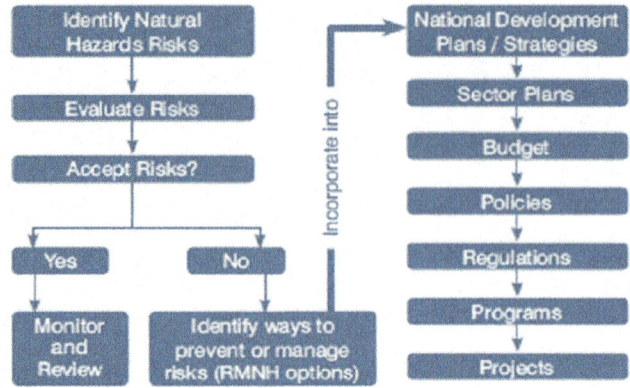

Fig. 9.3 Planning for disaster risks (Source: Bettencourt et al. 2006)

Linnerooth-Bayer et al. 2011) and their acceptability. Cost and benefits, in turn, depend on the nature of the hazard and risk. One way to think about the interaction and effectiveness of prevention and risk financing is by taking a layering approach shown in Fig. 9.4 (see also Chap. 16 in this book by Hochrainer-Stigler et al.).

For the low – to medium-sized loss events, which happen relatively frequently, prevention is likely to be more cost effective in reducing burdens. The reason is that the costs of prevention often increase disproportionately with the severity of the consequences. Moreover, individuals and governments are generally well able to finance lower consequence disaster events from their own means, for instance, savings or calamity reserve funds, and including international assistance. The opposite is generally the case for risk-financing instruments, including reserve funds, catastrophe bonds and contingent credit arrangements. These instruments do not reduce losses, but the variability around the losses, and thus only become cost-effective at higher costs associated with lower probability (e.g., at 100 year events). For this reason, it is generally advisable to use risk based instruments mainly for lower probability events that may have debilitating consequences (catastrophes). Finally, as shown in the uppermost layer of Fig. 9.4, individuals and governments will generally find it too costly to use risk financing (or risk reduction) instruments against very extreme risks occurring less frequently than, say, every 500 years. Overall, by taking a probabilistic approach, we can inform the full spectrum of options across the continuum of risk measures. As described below, our analysis which is based on the CATSIM model, provides an illustration of a probabilistic analysis.

9.4 Assessing and Planning for Economic Risk: CATSIM

When applying CATSIM to the Nepal case, we generally follow the approach discussed in Hochrainer-Stigler et al. (2013), but add a sectoral impact analysis module based on a Social Accounting Matrix (SAM) approach in order to represent

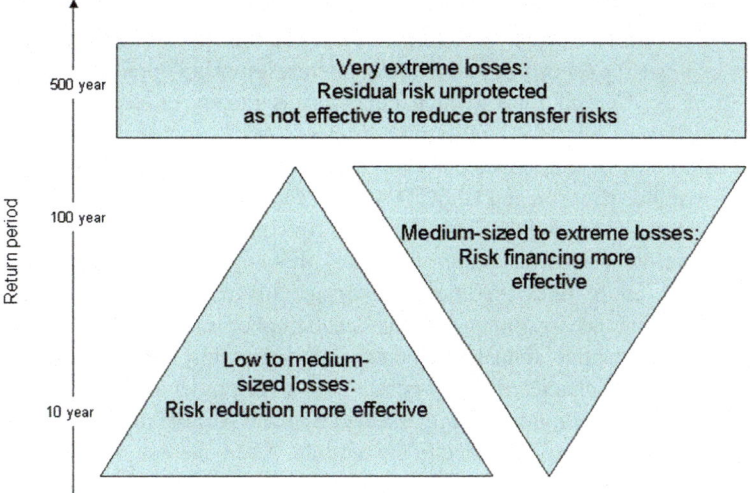

Fig. 9.4 The layering approach for risk reduction and risk financing

the distributional impacts of disasters. To properly introduce the analysis, we shortly describe the CATSIM methodology, which is organized around five steps, and we then give more emphasis to the analysis of aggregate and sectoral impacts as essential for step 4.

9.4.1 Methodological Steps of CATSIM

Step 1: Assessing asset risk
In the first step, risk is assessed in terms of the probability of asset losses (also called direct losses) in a relevant country or region. Consistent with general practice, risk is modeled as a function of hazard (frequency and intensity), the elements exposed to those hazards and their physical vulnerability. We assume that in the case of a disaster event, the government of Nepal will need to take responsibility for the following: (i) reconstruction of public assets: roads, bridges, schools, hospitals; (ii) provide relief and reconstruction support to private households and businesses temporarily affected; (iii) provide relief to the poor. This step in the methodology of CATSIM involves devising loss-frequency distributions, which relate probabilities to loss of assets.

Step 2: Assessing economic and fiscal resilience
A key aspect in CATSIM is the operationalization of economic and fiscal resilience. The focus is on the availability of internal and external savings of a country or region to spread risks and refinance losses as well as increased post-disaster expenditure needs, e.g. for supporting the private sector with relief and recovery assistance. Fiscal resilience is determined by the general conditions prevailing in an economy, i.e.,

changes in tax revenue have important implications on a country's financial capacity to deal with disaster losses. Governments can raise resources ex post or after a disaster by diverting funds from other budget items, imposing or raising taxes, taking credit from the Central Bank (which either prints money or depletes its foreign currency reserves), borrowing by issuing domestic bonds, borrowing from international institutions, issuing bonds on the international markets, and finally asking for outside assistance (Benson and Clay 2004; Fisher and Easterly 1990).

In addition to accessing ex post sources, a government can arrange for financial provisions before a disaster occurs (ex ante). Ex ante financing options include reserve funds, credit lines, traditional insurance instruments (public or private), or alternative insurance instruments, such as catastrophe bonds. These ex ante options can involve substantial annual payments and opportunity costs. Each of these financing sources is characterized by costs to the government as well as factors that constrain their availability, which are assessed by the CATSIM module. Sources not considered feasible are not included in the module. As an example, disaster taxes are expensive to administer and generally are not part of the public sector financing portfolio. Borrowing to finance deficits in the budget is heavily constrained by existing deficits and debt. To provide detail regarding the debt constraints, we employ a debt threshold for the present value of debt of 250% of revenue, which is often considered a critical debt threshold not to be exceeded (see HIPC 2002).

Step 3: Measuring financial and fiscal vulnerability by the "resource gap"
Using the information on direct risks and fiscal resilience, financial (or fiscal) vulnerability can be evaluated. Financial vulnerability is thus defined as the lack of access of a government to domestic and foreign savings for financing reconstruction investment and relief post-disaster. The shortfall in financing is measured by the term resource gap (or fiscal/financing gap). The resource gap is understood as the lack of financial resources to restore assets lost due to natural disasters and to continue with development as planned.

Step 4: Mainstreaming disaster risk into development planning
Ultimately the implications of disaster risk on economic development and other "flow variables" is of major interest when mainstreaming disaster risks into development planning and macroeconomic analysis. For that matter, direct risk, fiscal vulnerability and the prevalent economic conditions in Nepal are combined in order to derive an estimate of potential fiscal and macroeconomic impacts, such as in terms of GDP effects.

Step 5:Reducing risk and building resilience
CATSIM can illustrate the pros and cons of strategies for building economic and fiscal resilience using ex-ante financial instruments. Vulnerability and resilience are understood as properties of dynamic economic and social systems, which can adapt and manage shocks and surprises. As discussed, there are two broad types of risk based policy interventions: those that reduce the risks of disasters by reducing exposure and physical vulnerability, and those that build resilience of the responding agencies, e.g. by using financial risk management options. As CATSIM for the case of Nepal was used to examine and document disaster risk rather than test and examine options, for the present case we do not go into further detail on this model feature.

9.4.2 Modeling Aggregate Impacts

This analysis focusses on step 1 of CATSIM, or risk assessment, and in the following section we provide more detail regarding some technical aspects of modeling economic disaster risk. For the aggregate impacts, a production function approach is utilized for assessing GDP losses. As is standard practice in macroeconomics, a Cobb-Douglas type production function specification is used to project GDP based on inputs of capital and labor.

$$Y = AK^{\alpha}L^{\beta} \qquad (9.1)$$

where K is capital, L is labor, Y represents GDP, and α, β are the production coefficients. For Nepal, parameters of the production function are estimated using data on GDP, capital and total labor force from 1980 to 2004 as given by the World Bank Development Indicators (World Bank 2012). Table 9.4 shows the results for the estimation of these parameters.

In a next step, GDP is "shocked" by reducing capital by a fraction ΔK lost in a disaster event

$$\Delta Y = AK^{\alpha}L^{\beta} - A\Delta K^{\alpha}L^{\beta} \qquad (9.2)$$

The fraction of capital destroyed is determined by the asset loss distributions determined in Step 1. The decrease in capital stock in each event scenario can be obtained from random draws from the loss distribution used in CATSIM.

9.4.3 Modeling Sectoral Impacts

A disruption of one industrial sector can affect other economic agents through interdependencies within an economy. This is called a higher order effect. Disruption in one factory, for example, would lead to reduced orders for needed components. It would cause its suppliers to decrease their production and, in turn, to reduce their orders for inputs. This would continue up the supply chain. In a similar vein, the shutdown of a factory leads to decreases of demand for labor. This decreases the income of households, which in turn reduces final consumption of products leading to multiplier effects within the economy.

Table 9.4 Parameters of the aggregate production function for Nepal

	Coefficient	S.E	t value	P-value
Constant	4.64	2.02	2.295	0.0316
Capital	0.46	0.06	7.149	0.001
Labor	0.60	0.20	3.005	0.007

R square: 0.996

Multipliers and sectoral impacts can be estimated employing Input-output (I-O) analysis. I-O analysis is a standard method to study the interlinkages between economic sectors and the interconnectedness within an economy. An I-O table displays the flows of transaction within an economy. A Social Accounting Matrix (SAM) is an extension of an I-O model and additionally summarizes the distribution of income across certain types of households (see Stone 1961; Pyatt and Thorbecke 1976; Pyatt and Roe 1977). I-O and SAM analyses are common tools for the estimation of economic losses from natural disasters (see for example Okuyama 2008).

Figure 9.5 illustrates the logic of the SAM approach as employed in this study for Nepal. It starts from defining a so-called primary loss for each industry at first, which is the loss in output caused by the loss of capital stock discussed above and distributed over the economic sectors.

Based on the information of the primary loss, higher-order effects can be calculated using the SAM multiplier matrix. Because a SAM is a demand-driven model, the input data must be indicated in terms of change in demand. A change in demand is calculated by dividing the primary loss by the diagonal elements of the multiplier matrix (see Okuyama and Sahin 2009). Economic losses are then calculated by multiplying the demand change to the multiplier matrix.

For the Nepal analysis, we employ the Social Accounting Matrix as calibrated by Acharya (2007) for Nepal. Based on the loss distributions estimated with CATSIM and the aggregate GDP estimates presented above, we calculate sector specific losses and income impacts for household groups taking into account higher-order effects. The characteristics of the SAM approach for the given case are as follows (see also Appendix for the multiplier matrix): (i) Three economic sectors are considered, (ii) the inputs to production are capital, low-skilled labor and high-skilled labor; (iii) there are four population groups earning income: urban households, large rural households, small rural households, and landless rural households.

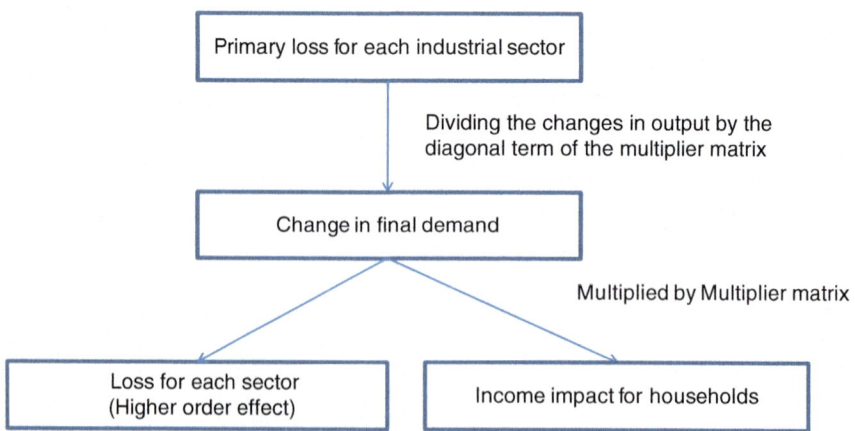

Fig. 9.5 Outline of the social accounting matrix approach

9.5   Results

We report results in terms of direct risk estimates, assessments of the economic risks as well as distributional impacts.

9.5.1   Assessment of Direct Asset Risks (Step 1)

In the first step, probabilistic asset risks for earthquake and flooding as well as the combined risks are estimated. Using information available on assets in Nepal as calculated for the project (see ADPC, NGI and IIASA 2010), a total value of capital stock of USD 75.3 billion is computed. As shown in Table 9.5, the values at risk for which the government is liable (contingent liabilities) are approximated overall at USD 37.7 billion. The calculation is made as follows: because little information is available on public sector capital stock in Nepal, it is assumed (in line with global averages) that approximately 30% of the total capital stock is public. Since one third of the population of Nepal is poor, the government will additionally absorb a large extra burden in the case of a disaster. Consistent with average figures (see Burby 1991; Freeman et al. 2002), it is further assumed that the government will have to spend an amount equivalent to another 20% of the total asset losses in order to provide relief.

The exposure information is combined with vulnerability and hazard data, and leads to individual loss distributions for flood and earthquake risk (see ADPC, NGI and IIASA 2010). These estimates are broadly based on the 1833 and 1934 earthquake events (estimated to have been 500 – and 100-year events, respectively) as well as 10-, 50 – and 100-year flood events (see Tables 9.6 and 9.7).

Table 9.5 Assets and government liabilities

Type	Billion USD
Private capital	52.8
Public capital	22.5
Total capital	75.3
Government contingent liabilities (public assets and assistance to private sector and households)	37.7

Table 9.6 Return periods and losses for flooding

Return Period	Probability	Losses(Million Nepali Rupee)	Losses(Million USD)
10	0.9	6,464	92
50	0.98	7,580	108
100	0.99	8,132	116

Table 9.7 Return periods and losses for earthquake risk

Return Period	Probability	Losses(Million Nepali Rupee)	Losses(Million USD)
100	0.99	1,017,827	14,540
500	0.998	1,102,685	15,752

Table 9.8 Potential joint losses associated with combined flood and earthquake risk

	Low estimate (Billion USD)	Central estimate (Billion USD)	High estimate (Billion USD)
20-year event loss	4.5	5.3	8.0
50-year event loss	7.0	9.7	10.5
100-year event loss	8.7	12.2	13.9
250-year event loss	12.0	16.8	17.6
500 year event loss	12.1	17.0	30.8

For our model-based economic risk assessment, a continuum of possible future risk scenarios is necessary, which requires fitting a whole distribution to these point estimates. Also, a convolution of flood and earthquake risk is required. Due to the small number of loss return periods available for the estimation of the extreme value distribution, past loss observations from the EM-DAT databases are also used to parametrize the distributions. Extreme value distributions were chosen from the broad class of the Generalized Extreme Value (GEV) distribution as well as the Generalized Pareto (GPD) and Weibull distributions. The GEV distribution performed best and was ultimately selected. This distribution is defined as:

$$H_{\xi(x)} = \begin{cases} \exp(-(1+\xi x)^{-1/\xi} & if\ \xi \neq 0 \\ \exp(-\exp(-x)) & if\ \xi = 0 \end{cases} \tag{9.3}$$

where $1 + \xi x > 0$

In order to account for the uncertainty associated with the estimation procedure, uncertainty bounds in the form of a central estimate (most probable case) as well as low and high estimates are examined. It is reasonable to assume that earthquakes and floods will occur independently, and we perform a convolution to get a joint loss distribution for Nepal. Convoluting two independent distributions with densities "f" and "g" is defined as:

$$(f * g)(x) = \int_{R^d} f(y)g(x-y)dy \tag{9.4}$$

We arrive at a joint distribution including low, central and high estimates with the following estimated losses (Table 9.8, Fig. 9.6).

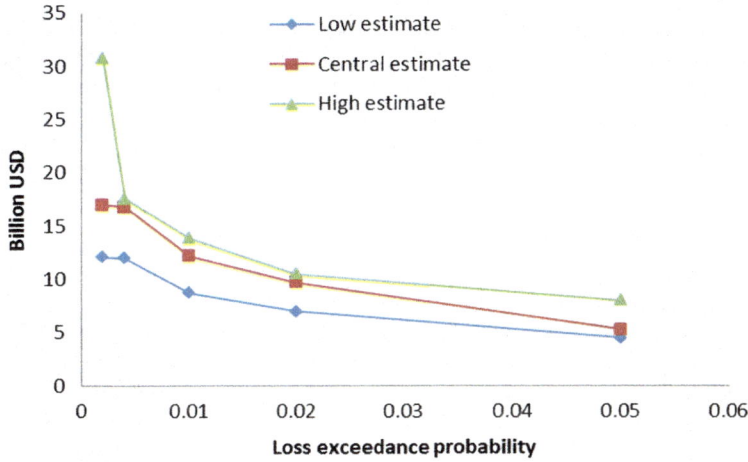

Fig. 9.6 Joint flood and earthquake risk distribution

9.5.2   Estimation of Fiscal Resilience of the Public Sector (Step 2)

An understanding of the sources for financing disasters in Nepal, including the costs and constraints, is crucial for planning a sound disaster risk management strategy. Concerning ex-post sources, Nepal is constrained by its fiscal inflexibility and low revenue base. Diversion from the budget is considered highly constrained, and therefore we assume that, as a maximum, 10% of the budget can be diverted. In line with empirical estimates across a sample of events, international assistance is assumed to be up to 10.4% of the total losses (see Freeman et al. 2002). Also, Nepal has very limited access to international capital markets, and thus it is assumed that Nepal can borrow only from multilateral sources at concessional rates and cannot issue any bonds in the international capital markets post disaster. The present value of external debt is over 240% of revenue in 2008. This means that the amount of debt which the government can additionally borrow from abroad is quite limited if a debt of 250% of revenue is considered a binding threshold for debt sustainability (see Table 9.9).

9.5.3   Fiscal Vulnerability and the "Resource Gap" (Step 3)

Summarizing all potential sources, the CATSIM model can provide an estimate of the government's fiscal vulnerability. It is most meaningful to assess the fiscal and economic consequences of exposure to both hazards jointly, as those are independent and thus may coincide in a short time span.

Table 9.9 Sources for financing disaster losses in Nepal

Source	Parameter value used
International donor assistance	10.4%
Diversion from budget	10%
Domestic bonds and credit	0
Multilateral borrowing	Very limited
Reserve fund	0
International borrowing	0

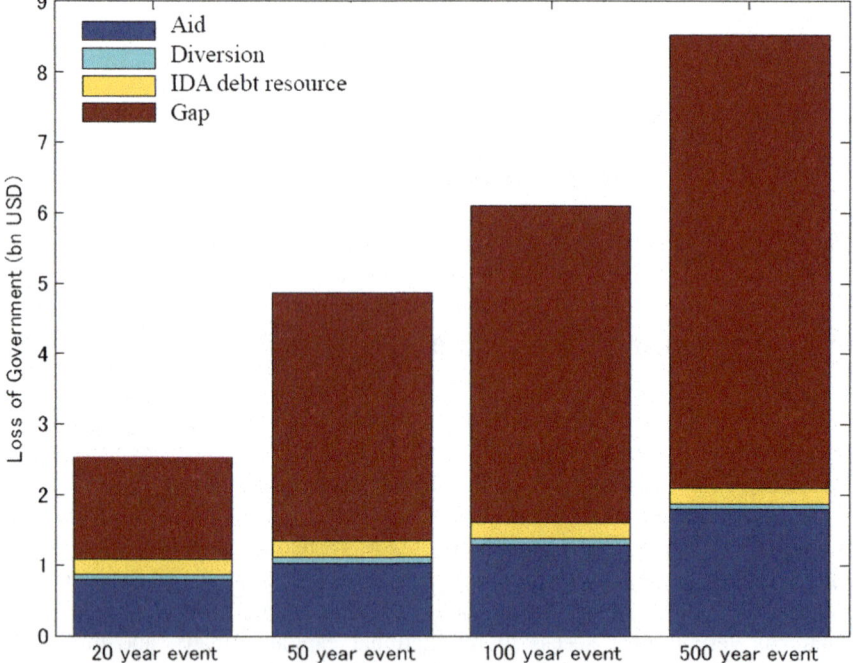

Fig. 9.7 Fiscal vulnerability and resource gap for the joint risk distribution (central estimate)

Given the assumptions and data as described above, Nepal's fiscal vulnerability to the combined risk is shown in Fig. 9.7 for the central estimate (we do not further show results for the low and high estimates in order to simplify the discussions and results generally are similar for these uncertainty ranges).

Fiscal vulnerability is considerably high and already for an event as frequent as a 20-year return period, the public authorities in Nepal would face difficulties raising sufficient funding. Here, the resource gap could amount to more than USD 1.3 billion, given our analysis. Available funding for this event includes aid inflows, which could amount to as much as USD 800 million, USD 50 million may be diverted from the budget, and another USD 150 million could be borrowed on highly concessional terms, such as offered by the World Bank through the International Development Bank (IDA). Keeping data limitations and restrictive

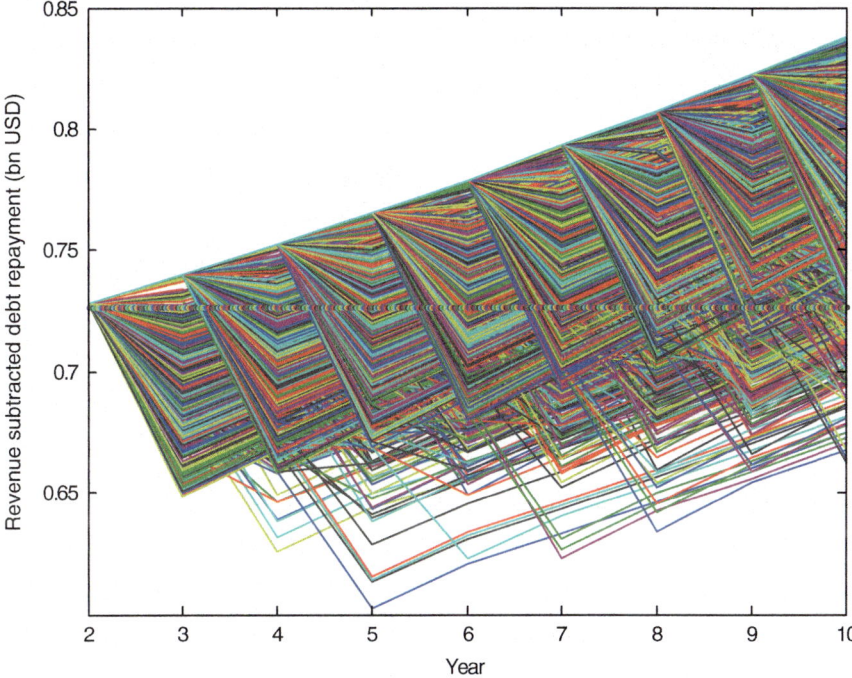

Fig. 9.8 Potential fiscal impacts due to the joint risk of flood and earthquake

assumptions in mind, this analysis shows that the government of Nepal has vastly insufficient financing resources available for tackling a relatively frequent 20-year event. The key driver behind the losses is earthquake risk, which is almost two magnitudes higher in terms of losses than flood risk.

9.5.4 Mainstreaming Disaster Risk into Macroeconomic and Development Planning (Step 4)

As a next step, risk and fiscal vulnerability is integrated with a model of the economic system in order to assess aggregate fiscal and economic effects.

9.5.4.1 Aggregate Analysis

As discussed, in order to mainstream and assess risk, a production function is established relating assets to flow variables (consequences of business interruptions etc.). Now using stochastic sampling approaches, fiscal and macroeconomic projections taking account of risk can be performed. Figures 9.8 and 9.9 show a

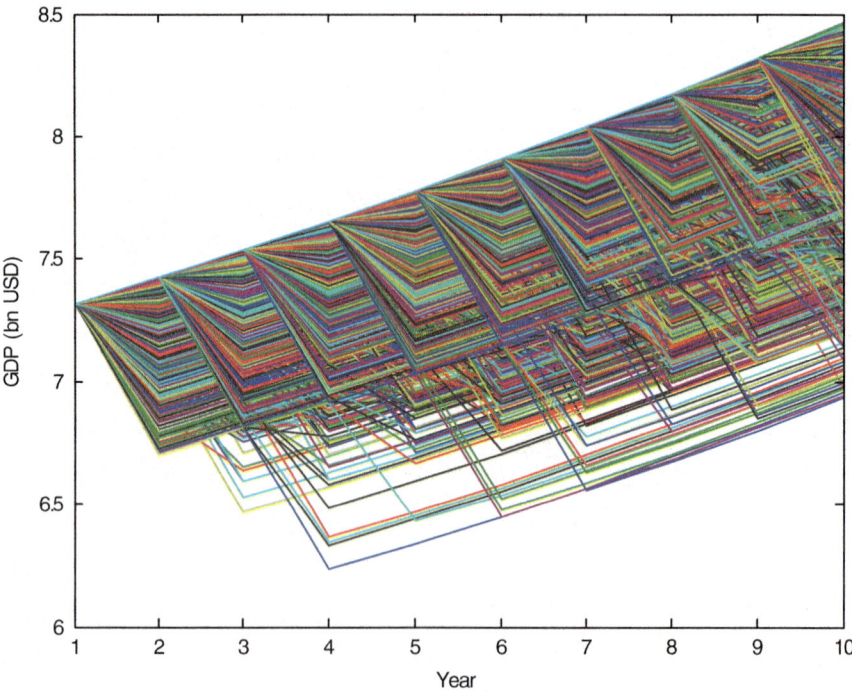

Fig. 9.9 Potential GDP impacts due to joint risk of flood and earthquake. Note: the time period considered is 2011–2020

selection of trajectories for fiscal and macroeconomic impacts for Nepal.[1] In Fig. 9.8, potential trajectories for government revenue (less repayments for debt) are outlined. This variable is a useful indicator representing budget flexibility (*fiscal space*) after mainstreaming disaster losses and government relief requirements into the projections. The graph shows that in the cases without disasters, budget flexibility would increase; yet in many instance, there is a potential for disasters seriously affecting budget flexibility, which in a number of scenarios would be highly reduced over the 10 year modeling time horizon.

We can further investigate fiscal space under disaster risk using the present value of government resources, discounted over time, and compare this to a baseline case without including disaster risk (see Table 9.10). In year 10, fiscal space as measured by the present value of budgetary resources would decrease on average by about 30% when explicitly factoring in disaster risk, and the standard deviation would equal about 36%. Considering another indicator for fiscal risk, the probability of a

---

[1] In the calculation, it is assumed that the private sector invests a certain ratio of GDP to capital if no disaster happens. If a disaster happens, the private sector does not get external funding for recovery (as it does not have access to the financial markets), so that the damaged capital cannot be restored immediately.

Table 9.10 Indicators for fiscal risk (in year 10 of the modeling time horizon)

Key variable	Decrease in PV of budget revenue (mean)	Decrease in PV of budget revenue (standard deviation)	Probability of resource gap
Joint earthquake and flood (central estimate)	29.7%	36.2%	57.8%

Note: PV is present value

resource gap, which is estimated at close to 60%, it seems very likely that events may occur in the near future that deteriorate public finances and cause longer term adverse macroeconomic impacts.

In a next step, aggregate economic risk is modeled and Fig. 9.9 identifies aggregate impacts on GDP based on risk, very limited resilience of the private sector and limited resilience of public authorities to respond. The GDP indicator shows that given limited resilience, disaster events may result in a lower economic growth trajectory. As with the fiscal impacts, the occurrence of such trajectories is stochastic and depends on the probability distribution of the losses (about 10,000 trajectories are calculated in this analysis). In a number of cases, GDP would be significantly reduced compared to the business-as-usual case.

9.5.4.2 Intersectoral Linkages

In a next step, we calculate the intersectoral impacts of disaster risk. Due to computational complexities involved in running the stochastic scenarios, the SAM approach cannot easily be reconciled with the risk analytical methodology, and we thus focus on a scenario earthquake event with a 100-year return period, which is roughly equal in intensity to the devastating event of 1934. The disastrous event of 1934, the so called Bihar/Nepal Earthquake, heavily affected the capital region of the Kathmandu Valley, killed about 9,000 people, destroyed 20% of the valley's assets and severely damaged another 40%. For such an event, asset losses of about USD 14.5 billion have been estimated for today as documented in Table 9.2. The primary affected sectors are housing, education, health, transportation, industry (manufacturing), and power infrastructure. Among them, the shutdown of the manufacturing sector would most seriously decrease its purchases of intermediate input. This study, therefore, focuses on the ripple effect due to a shutdown of the manufacturing sector. Table 9.11 and Fig. 9.10 summarize the primary loss and calculated loss as well as income impact of households for this scenario earthquake as one example. It can be observed that the primary GDP loss (about USD 730 million) is doubled (about USD 1,420 million) by the multiplier effect as economic interdependencies reduce demand for agricultural goods as well as commercial and public services. The total value of the higher order loss would thus amount to as much as approximately 19% of current GDP, which seems reasonable for such a catastrophic event destroying a fifth of the total assets in Nepal.

Table 9.11 Primary and higher order losses of a scenario earthquake of the severity of the 1934 earthquake (current million USD)

Sector	Primary GDP loss	Higher order GDP loss	Income loss
Agriculture	–	383.3	–
Industry	731.5	731.5	–
Commercial service	–	228.6	–
Public service	–	80.7	–
Urban household	–	–	201.4
Large rural household	–	–	143.0
Small Rural household	–	–	181.9
Landless rural household	–	–	97.3
Total	731.5	1,424.1	624.5
% GDP	10%	19%	

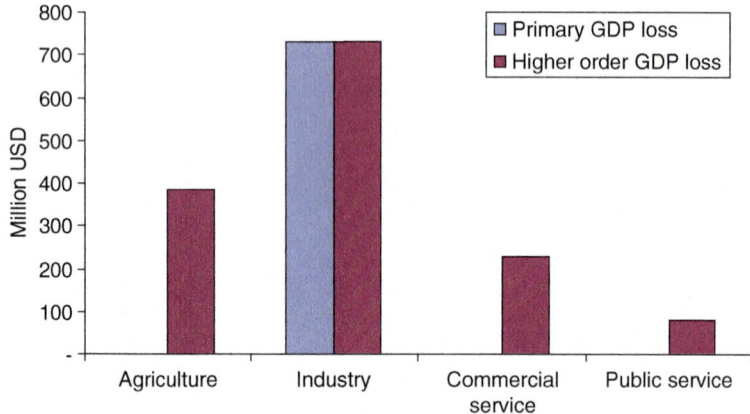

Fig. 9.10 Primary and higher order losses for a 1934 scenario earthquake (in absolute USD terms)

As to the distribution of the shock, income losses may spread from urban areas to the countryside due to reduced demand for agricultural commodities by those directly affected. Consequently income for rural households, which are not directly affected by such an event, would ultimately also be reduced (Fig. 9.11).

Our analysis underlines the potentially large distributional effects of a catastrophe event, such as a 100-year earthquake, on overall GDP as well as associated linkage effects on the different households considered. It is worth noting that it would be desirable to crosscheck our results with recorded economic impacts associated with large disasters in Nepal, particularly impacts for the 1934 catastrophe. Yet, as outlined in the introduction, these effects are often hidden, and for the 1934 event, beyond discussions regarding the immense suffering, there is no public

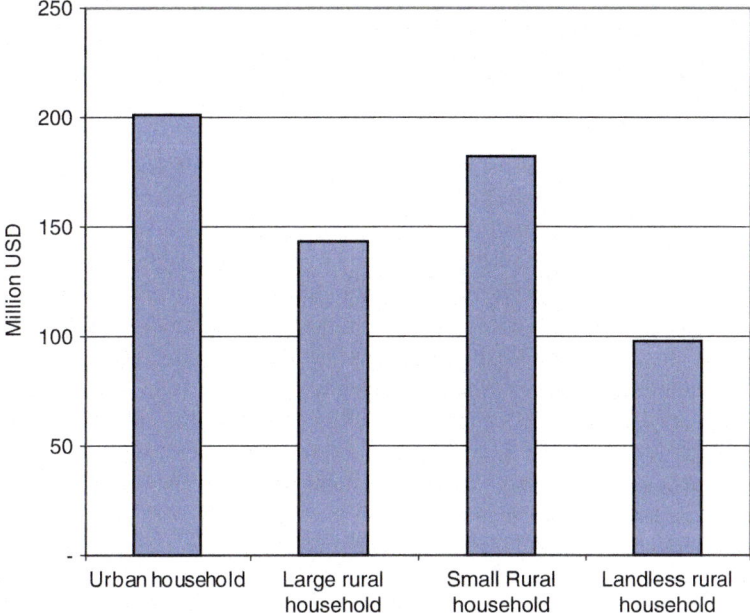

Fig. 9.11 Income effects for an earthquake of the size of the 1934 event

record of the economic effects. Also, it has to be noted here, that this analysis is of short term nature (1 year) and in the medium term response mechanisms facilitating a recovery would need to be considered. Using the CATSIM framework, risk and disaster management options may be identified and tested in this modeling framework, which, however, was not the purpose of the project, and had to be left as a potential consideration for further model applications.

9.6   Discussion and Conclusions

Disasters are considered a serious threat to lives and property in Nepal. The disaster burden imposed is considered heavy, yet little is known regarding the true economic impacts and losses. For this matter, economic risk analysis involving catastrophe modeling was undertaken and formed an important part of the project on developing an all-hazards model for Nepal.

The risk-based economic analysis reported here centered on examining the fiscal and economic effects of earthquake and flood risk across Nepal, which are often considered the key hazards leading to macroeconomic impacts. The analysis shows that the economic and fiscal risks posed by natural disasters are large for Nepal, and there is a clear case for considering these impacts in economic and fiscal planning. In particular, earthquake risk, for which a 100-year event of the size of the 1934 event may mean losses of about USD 15 billion, can lead to large fiscal and economic impacts. Yet, given limited fiscal resilience, an event as frequent as every 20 years may already lead to a resource gap, which is defined as the inability to provide key relief and reconstruction requirements post disaster. If disaster risk is explicitly considered in economic planning and a 10-year time horizon, we find fiscal space reduced by about 30% compared to a case without consideration of disaster risk. Finally, when using a social accounting matrix approach to derive intersectoral linkages, we estimate that a large disaster event, such as that of the size of the 1934 earthquake would lead to substantial (in the range of 20%) reductions in GDP due to linkages across primarily unaffected sectors, such as agriculture.

As a key application, the economic modeling presented may inform contingency liability planning in disaster exposed and vulnerable countries. As another application, CATSIM modeling may inform relief and reconstruction efforts post event. The analysis demonstrates that disasters like earthquakes and floods may ripple through an economy and indirectly affect sectors that were not impacted directly by the disaster event. As a consequence, it is worth considering important cross-sector linkages in any strategy that touches upon the consequences of disaster risk on the economy overall.

We conclude that in the face of massive disaster risk coupled with limited resilience, the Nepalese government should consider a position of risk aversion, and risks should be explicitly accounted for (*ex ante* approach). The need for taking a risk averse position can also be supported by looking at the empirical indicators related to disaster spending shown in Sect. 9.2. Even in the absence of major disasters, the government of Nepal is already highly dependent on foreign support, with about a third of the country's income from foreign aid and more than 60% of development spending financed by multilaterals. As disasters and the necessary spending for relief and reconstruction often lead to a significant loss of these scarce resources, implementing options for limiting such a drain on funding is important.

Finally, a word of caution has to be expressed regarding model calibration as well as the risk estimates. The results are necessarily associated with considerable uncertainties, which were captured by us where possible. For example, we used mean estimates and also incorporated uncertainty in terms of estimates of low and high risk. These often large uncertainties need to be factored into any analysis, and before attempting to derive very specific policy recommendations for risk management, there is a need for further investigations and discussion with key experts and stakeholders on policy priorities including disaster risk options. In Nepal, a process involving the Ministry of Home Affairs, donors and development banks has been initiated. We hope that the modeling presented here may ultimately contribute to better bolstering Nepal against disaster risk.

Appendix

SAM Multiplier Coefficients Table for Nepal Based on Acharya (2007)

	Agriculture	Industry	Commerce	Public sector	Wage to low-skilled labor	Wage to high-skilled labor	Capital	Urban household	Large rural household	Small rural household	Landless rural household	Firms
Agriculture	1.94	0.78	0.72	0.62	1.13	1.03	0.89	1.05	0.74	1.22	1.33	0
Industry	0.43	1.50	0.42	0.41	0.58	0.54	0.47	0.53	0.43	0.67	0.60	0
Commerce	0.64	0.47	1.65	0.53	0.74	0.74	0.63	0.90	0.54	0.71	0.72	0
Public sector	0.20	0.17	0.22	1.17	0.27	0.25	0.21	0.26	0.19	0.26	0.33	0
Wage to low-skilled labor	0.74	0.44	0.54	0.56	1.54	0.51	0.44	0.54	0.37	0.57	0.62	0
Wage to high-skilled labor	0.20	0.19	0.18	0.29	0.18	1.17	0.14	0.18	0.13	0.19	0.20	0
Capital	0.96	0.75	0.96	0.52	0.77	0.73	1.63	0.80	0.54	0.81	0.84	0
Urban household	0.56	0.41	0.50	0.42	0.71	0.81	0.66	1.45	0.31	0.46	0.49	0
Large rural household	0.39	0.29	0.36	0.28	0.45	0.57	0.49	0.32	1.22	0.33	0.34	0
Small rural household	0.53	0.37	0.45	0.38	0.76	0.60	0.57	0.42	0.28	1.43	0.46	0
Landless rural household	0.29	0.20	0.24	0.22	0.47	0.33	0.27	0.23	0.16	0.24	1.25	0
Firms	0.13	0.10	0.13	0.07	0.11	0.10	0.22	0.11	0.08	0.11	0.12	1

References

Acharya S (2007) Flow structure in Nepal and the benefit to the poor. Econ Bull 15(17):1–14

ADPC, NGI, IIASA (2010) Nepal all hazards risk assessment. Project Report to the Government of Nepal by the Asian Disaster Preparedness Center, Norwegian Geotechnical Institute and International Institute for Applied Systems Analysis, Bangkok

Benson C, Clay E (2004) Understanding the economic and financial impacts of natural disasters, Disaster risk management series no. 4. World Bank, Washington, DC

Bettencourt S, Croad R, Freeman P, Hay J, Jones R, King P, Lal P, Mearns A, Miller G, Pswarayi-Riddihough I, Simpson A, Teuatabo N, Trotz U, Van Aalst M (2006) Not if but when – adapting to natural hazards in the Pacific islands region: a policy note. East Asia and Pacific Region, Pacific Islands Country Management Unit. World Bank, Washington, DC

Bhattarai D, Chhetri MB (2001) Mitigation and management of floods in Nepal. Ministry of Home Affairs, HMG/Nepal

Burby JR (1991) Sharing environmental risks: how to control governments losses in natural disasters. Westview Press, Boulder

CRED (2010) Centre for research on the epidemiology of disasters international disaster database. www.em-dat.net. Université Catholique de Louvain, Brussels, Belgium

Fisher S, Easterly W (1990) The economics of the government budget constraint. World Bank Res Obs 5(2):127–142

Freeman PK, Martin LA, Linnerooth-Bayer J, Mechler R, Saldana S, Warner K, Pflug G (2002) Financing reconstruction. Phase II background study for the inter-American development bank regional policy dialogue on national systems for comprehensive disaster management. InterAmerican Development Bank, Washington, DC

HIPC (2002) About the HIPC initiative, Washington, DC. http://www.worldbank.org/hipc/about/hipcbr/hipcbr.htm

Hochrainer-Stigler S, Mechler R and Pflug G (2013) Modeling macro scale disaster risk: the CATSIM model. In: Amendola A, Ermolieva T, Linnerooth-Bayer J, Mechler, R (eds) Integrated catastrophe risk modeling: supporting policy processes, Springer, Dordrecht, pp xx–xx

Linnerooth-Bayer J, Mechler R, Hochrainer S (2011) Insurance against losses from natural disasters in developing countries. Evidence, gaps and the way forward. J Integr Disaster Risk Manag 29(1):57–82

Okuyama Y (2008) Critical review of methodologies on disaster impacts estimation. Background paper for World Bank report economics of disaster risk reduction. World Bank, Washington, DC

Okuyama Y, Sahin S (2009) Impact estimation of disasters: a global aggregate for 1960 to 2007. Policy research working paper 4963, World Bank, Washington, DC

Pyatt G, Roe AR (1977) Social accounting for development planning. Harvard University Press, Cambridge

Pyatt G, Thorbecke E (1976) Planning techniques for a better future. International Labor Organization, Geneva

Stone JRN (1961) Input-output and national accounts. Organization for Economic Cooperation and Development, Paris

Upreti S (2006) The nexus between natural disasters and development: key policy issues in meeting the millennium development goals and poverty alleviation. Economic policy network, Policy paper 27. Government of Nepal and Asian Development Bank, Nepal "Resident Mission"

World Bank (2009) Nepal at a glance. World Bank, Washington, DC

World Bank (2012) World bank development indicators. World Bank, Washington, DC

Part III
Tisza River Basin in Hungary: Flood Risk Management, Multi-stakeholder Processes and Conflict Resolution

Dedicated to the memory of Zsuzsanna Flachner

Chapter 10
Catastrophe Models and Policy Processes: Managing Flood Risk in the Hungarian Tisza River Basin – An Introduction

Joanne Linnerooth-Bayer, Love Ekenberg, and Anna Vári

Abstract In this chapter we provide an introduction to this section of six chapters, which examine how catastrophe models can contribute insights to multi-stake-holder policy processes by focusing on flood risk management in the Hungarian reach of the Upper Tisza river. The flood problem in this vulnerable region remains today acute mainly because of increasing flood risk due primarily to land use and perhaps also to climate change, as well as to a management regime in flux. A recent popular movement to change the management regime from the traditional *river defense paradigm* (RDP) to a more environmentally oriented *working landscape paradigm* (WLP) has been stalled. This stalled regime shift highlights the critical importance of reaching consensus, not only on flood measures that promote the sustainable development of the region, but also on the distribution of the losses from floods. The papers in this section focus on the latter by demonstrating how catastrophe models can aid a participatory process aimed to design a flood insurance and public compensation system. In addition, the papers address flood risk in the region, and how it will be impacted by climate change, as well as the vulnerability of the Tisza basin residents.

Keywords Multi-stakeholders policy processes • Insurance program • Catastrophe modeling • Tisza river basin • Flood risk management • Climate change impact

J. Linnerooth-Bayer (✉)
Risk, Policy and Vulnerability (RPV) Program, International Institute for Applied Systems Analysis (IIASA), Schlossplatz 1, A-2361 Laxenburg, Austria
e-mail: bayer@iiasa.ac.at

L. Ekenberg
Department of Computer and Systems Sciences, Stockholm University, Stockholm, Sweden

A. Vári
Institute of Sociology, Hungarian Academy of Sciences, Budapest, Hungary

A. Amendola et al. (eds.), *Integrated Catastrophe Risk Modeling: Supporting Policy Processes*, Advances in Natural and Technological Hazards Research 32,
DOI 10.1007/978-94-007-2226-2_10, © Springer Science+Business Media Dordrecht 2013

10.1 Scope

This section of six chapters examines how catastrophe models can contribute insights to multi-stakeholder policy processes by focusing on flood risk management in the Hungarian reach of the Upper Tisza river. The flood problem in this vulnerable region remains today acute mainly because of increasing flood risk due primarily to land use and perhaps also to climate change, as well as to a management regime in flux. A recent popular movement to change the management regime from the traditional *river defense paradigm* (RDP) to a more environmentally oriented *working landscape paradigm* (WLP) has been stalled. The reasons are complex, including major cost overruns in the construction of the necessary reservoirs, and also a lack of political will especially for subsidizing changed land use practices (Borsos B, 2011, UNDP Technical Advisor, personal communication, 10 February, Budapest). Adding to this list are burden-sharing issues. In planning a reservoir in the Bereg region, for example, farmers demanded far more for their expropriated land than the government was offering (Borsos B, 2011, UNDP Technical Advisor, personal communication, 10 February, Budapest). This stalled regime shift highlights the critical importance of reaching consensus, not only on flood measures that promote the sustainable development of the region, but also on the distribution of the costs and benefits of these measures (Sendzimir et al. 2010).

This set of chapters shows how modeling research can contribute to flood risk management and particularly to the exploration and resolution of conflicts on who gains and who loses from social and environmental policies. For the most part, this research was conceived and carried out before recent attempts to shift the Tisza flood risk management to more environmentally oriented pathways; yet, the methods and results are highly relevant to current events. After examining present-day vulnerability in the region, the chapters turn to the design of a country-wide flood compensation and insurance system, and show how catastrophe models have helped to clarify the distributional issues inherent in any risk-sharing system. These latter chapters are based on interviews and workshops with stakeholders in the Upper Tisza region that were part of a project funded by The Swedish Research Council for Environment, Agricultural Sciences and Spatial Planning, and carried out by IIASA, Stockholm University and the Hungarian Academy of Sciences.

Given the critical importance of institutions and procedures for sharing benefits and costs, it is surprising that the discourse on the recent regime shift in the Tisza region has not confronted this issue in a rigorous manner – perhaps leading in some part to the delays we are currently witnessing. It is hoped that the methodologies presented in this section, many of which were designed to aid participatory processes, will be useful for all research intent on providing useful information to controversial policy issues such as those characterizing the policy discourse today in Hungary.

We begin by describing the history and current policy debates surrounding the controversial regime shift from flood defense to ecological and adaptive management of the river, and argue that risk sharing systems (compensation and insurance) are vital for enabling such shifts. We then turn to describing the chapters in this section, and conclude by underlining the importance of integrated research that connects the economic, social and ecological systems for sustainable management of today's river basins.

10.2 Background and Context

With its origins in the Ukrainian Carpathian mountains, the Tisza river flows along the border of Romania and then southwest across the great Hungarian plain, eventually flowing into the Danube river in the Serbian Republic. Precipitation falling on a vast mountainous area is concentrated in the Tisza, resulting in some of the most sudden (24–36 h) and extreme (up to 12 m) water floods in Europe (Halcrow Group 1999; Koncsos and Balogh 2007). Such extremes occur on average every 10–12 years in the Tisza basin (Bran and Borsos 2009), but the last century has seen rising trends in all facets of flooding: flood peak height, volume and frequency. The Tisza ranks as one of the highest flood-risk areas, as well as one of the poorest, in Europe.

The historical response to flooding has been in the form of hydro-engineering operations that have reconfigured the river basin (Balogh 2002). In the early eighteenth century the Tisza region was a diverse landscape with ploughed land, forests, floodplain orchards, meadows, fishing and cattle that co-existed with frequent and routine flooding (Andrásfalvy 1973; Bellon 2004). To provide the conditions for large-scale intensive agriculture (mainly wheat) and transport, the river was canalized, straightened and bracketed with levees to prevent flooding, and the floodplains were drained. As an unintended result, sediments previously flushed out by floods accumulated on the floodplain. Complaints by local farmers eventually provoked an agreement to expropriate the entire floodplain, which had previously been common land, for grain production.

10.2.1 The River Defense Paradigm

In recent history, the Hungarian government has continued to invest huge sums in the vast network of protective levees, including about 3,000 km of embankments, which must be continually heightened to protect against increasingly worsening floods (Balogh 2002). The government also typically takes responsibility for private damages in the event of a levee breach, compensating victims generously for groundwater inundation and other types of flood damage. After the Tisza floods in 2001, for instance, the government fully rebuilt nearly 1,000 houses that had been washed away (Vári et al. 2003).

Estimates show that damage to built capital and commerce from a major flood event could reach as high as 25% of the basin's GDP, or 7–9% of national GDP (Halcrow Group 1999). Understandably, the government is concerned about its tradition of taking almost full responsibility for flood risk management, including flood prevention, response, relief and reconstruction. Hungarian membership in the European Union has committed the government to a program of fiscal austerity, and for this reason the financial authorities would welcome more private responsibility for the reduction and response to flood disasters.

Moreover, critics of the "river defense paradigm" argue that it is neither economically nor ecologically sustainable especially in light of climate change (Sendzimir et al. 2010; Werners et al. 2010a, b). The levees remain inadequate to protect against increasing frequency and discharges of floods (Balogh 2002), and the lack of water retention is aggravating another problem in the region, scarcity of adequately clean water resources in dry periods. Ecologically, the RDP threatens existing unique freshwater wetland ecosystems that can provide valuable eco services and biodiversity so essential if the region switches to a more diversified agricultural and livelihood production (Sendzimir and Flachner 2007). This switch, many argue, is necessary given the increasing economic, environmental and social impoverishment of the region.

10.2.2   The New Vasarhelyi Plan

In response to the decline of the Tisza region, a network of experts, NGOs and intellectuals (called the "shadow network") recently became influential in promoting a changed management paradigm (Sendzimir et al. 2010). Instead of flood defense strategies implemented from Budapest in pursuit of building profitable export agriculture in the floodplain, this network advocated policies that enhance biodiversity, restore ecosystems and produce a more diversified landscape of livelihoods in the region. This would mean a shift from the *flood defense paradigm* to a *working landscape paradigm*.

Supported by the European Union, the shadow network organized a broad participatory process that gained the attention of the Hungarian government, which itself was facing huge costs from it flood defense policies. This social movement, along with four extreme flood events revealing the insufficiency of the levees, prompted the launching in 2003 of the New Vasarhelyi Plan (VTT) that emphasizes environmental protection and nature conservation (Government Decision No. 1107/2003 (X1.5). The new strategy calls for (i) reinforcing dikes where they do not meet the once-in-a-century standards required by the EU Water Directive, (ii) improving flood conveyance of rivers (reducing summer dikes, rehabilitating pastures and mosaic-type forests, and (iii) increasing existing and creating new flood plain areas, i.e., providing enough room for the river. For the latter, 75,000 ha of detention basins have been selected with a storage capacity of 1.5 b m^3 (about 6% of basin annual runoff), which engineers predict should be enough to decrease peak level events by 1 m all along the Tisza.

Views differ on the success of the VTT. While proponents claim that in its first phase (2003–7), 6 out of 11 retention basins would be scheduled for restoration (Bran and Borsos 2009), critics claim that only one new retention basin, the Ciga'nd

Polder covering about 25 km^2 of floodplain, has been built, and that little has been invested in rehabilitating pastures and forests. Because of cost overruns and an increasingly unfavorable economic environment, the planned reservoirs were reduced from 11 to 6. According to these critics, the failure of the regime shift called for by the VTT can be attributed to complex and systemic institutional and procedural issues (for a full description, see Sendzimir et al. 2010), of which the distribution of the burdens from its implementation appear to have played a role along with many other factors including escalating costs and lack of political will outside the water authorities (Borsos B, 2011, UNDP Technical Advisor, personal communication, 10 February, Budapest).

In hindsight, it appears that too little attention may have been given to the distribution of the costs of implementing the VTT. While the VTT contains clauses about compensating farmers for inundated lands and crop losses, there have been no estimates of what this would mean for the government's budget. Calls for compensation have greatly exceeded what the government can reasonably afford, which may be one factor leading to a breakdown in the government's resolve to implement the plan.

10.2.3 Sharing Costs: A Hungarian Compensation and Insurance System

Concurrent with the unsatisfactorily slowed implementation of the VTT, at least as perceived by many of its proponents, another related policy arena was experiencing similar difficulties in implementing government policy. The Hungarian government has recently legislated a nation-wide flood insurance program with the aim of shifting much of the post-flood burden from the government's budget to accumulated funds in this program (Linnerooth-Bayer et al. 2006). The insurance was fully underwritten by the government, and very low-income households would receive subsidies to enable them to purchase policies. However, insurance uptake remains extremely low because of the unwillingness of Hungarians to pay the premiums, and the unwillingness of insurers to write policies covering damages except those from breached levees. It is perhaps not coincidental that in the same year as the launch of the VTT, the government also launched its Wesselenyi fund (guaranteed by the state) for the compensation of uninsured damages caused by water.

10.3 How Models Can Contribute to Flood Risk Management

This book is dedicated to the notion that the development of efficient and equitable policies for managing disaster risks and adapting to global environmental change is critically dependent on robust decisions supported by integrated modeling. The chapters in this section examine how catastrophe models can provide insights on flood risk management in the Tisza region with the intent of contributing to policy processes – and robust decisions – that reduce risk and vulnerability of the mainly low-income residents.

We begin, not with a chapter on catastrophe modeling, but an empirical analysis titled *Social Indicators of Vulnerability to floods*, by Anna Vári, Zoltan Ferencz and Stefan Hochrainer. The analysis is based on an empirical survey conducted in the Bodrogköz and in the Bereg region within the Tisza flood basin. The questionnaire revealed that, while impacts are dependent mainly on exposure (location), important factors influencing vulnerability included: health status, education, savings, availability of post-flood financing, trust in the community and its institutions, and preparedness of institutions. Setting the stage for the chapters that follow, among other recommendations the authors note the importance of access to loans and other routes for obtaining post-disaster financing.

The following chapters point to alternative designs for a flood insurance program that are based on stakeholder views, and importantly aided by a flood model of the region. They underline the importance of identifying realistic flood management strategies considered fair by the stakeholders in the region and elsewhere. A main issue was to investigate different insurance schemes in combination with governmental compensation, or combining private responsibility with nation-wide solidarity. The research was focused on the Palad-Csecsei basin (the pilot basin), which is situated in the Szabolcs-Szatmár-Bereg County in northeastern Hungary. This region is one of the poorest agricultural regions of Europe, and floods repeatedly strike large areas.

The chapter by Joanne Linnerooth-Bayer, Anna Vári, and Lisa Brouwers shows how a participatory process can be aided by a computer model. Their chapter titled *A Model-Based Stakeholder Approach* for *Designing a Flood Management and Insurance System in Hungary* takes account of contending views on the flood problem and solutions held by the Hungarian stakeholders, including the public, the local authorities, government ministries and private insurers. The challenge was to design a national flood insurance system that would provide incentives for reducing flood risk, as well as fairly compensating victims in this poor and vulnerable region. A 3-year stakeholder process was aided by a catastrophe model that helped to clarify the distributional issues by showing how simulated flood losses would be shared among the victims, the government and the insurers depending on the design of the insurance pool.

This simulation model and its use in the Tisza participatory process is described in more detail in a chapter titled *Consensus by Simulation: a Flood Model for Participatory Policy* by Lisa Brouwers and Mona Riabacke. This chapter gives important details on the design, implementation and use of the dynamic and spatially explicit flood simulation model, which incorporated micro-level representation and Monte Carlo techniques. The model was equipped with an interactive graphical user interface designed for the particular context to facilitate its use as a decision support tool in the participatory setting with multiple users. The model supports comparisons between pre-defined policy options as well as the design of new policy options. During the concluding workshop the model was used interactively by the stakeholders to aid their decision making process.

In the absence of a stakeholder process, consensus can be modeled as shown in an innovative chapter by Mats Danielson and Love Ekenberg, titled *A Risk-Based Decision Analytic Approach to Assessing Multi-Stakeholder Policy Problems.* With an understanding of the preferences of the stakeholder groups, the authors show that decision analysis can be a useful tool in establishing and ranking different policy alternatives. The approach was employed to assess options for designing a public-private insurance and reinsurance system in the Tisza case. The design of a nation-wide insurance system involves handling imprecise information, including estimates of the stakeholders' utilities, outcome probabilities, and other complications, all of which this chapter addresses. The general method of probabilistic, multi-stakeholder analysis extends the use of utility functions for supporting evaluation of imprecise and uncertain facts.

Using a different methodology Tatiana Ermolieva, Yuri Ermoliev and Istvan Galambos show how stochastic optimization can help in the design of a national insurance pool. In their chapter titled *Financial Instruments in Integrated Catastrophic Flood Management: Demand for Contingent Credits* they develop a flood management model that takes into account the inherent complexities in catastrophic risk management: highly mutually dependent flood losses, the lack of information, the need for long term perspectives and geographically explicit analyses as well as the involvement of various agents such as individuals, governments, insurers, reinsurers, and investors. Making realistic assumptions on the preferences of these groups, the authors design an "optimal" public multi-pillar program involving partial compensation to flood victims by the central government, the pooling of risks through a mandatory public insurance on the basis of location specific exposures, and a contingent ex-ante credit to reinsure the pool's liabilities. Policy analysis is guided by the GIS-based catastrophe models and stochastic optimization methods with respect to location specific risk exposures.

In the final chapter in this section, *Flood Loss Considerations and Adaptation Strategies due to Climate Change in Hungary and the Tisza Region,* Stefan Hochrainer, Reinhard Mechler, Nicola Lugeri and Georg Pflug address climate change and its implications for the Tisza region. Many regions and sectors in Europe are vulnerable to increasing disaster risks and climate adaptation is moving to the forefront of EU and national policy. Yet, little is known about changing risks and possible adaptation options under dynamic conditions. The Tisza region is one of the hot spots in Europe and a prime case to study new risk assessment methods and risk management techniques in light of a changing climate. Based on a risk modeling approach the authors present indicative quantitative results on the part of climate change on future flood losses for Hungary with a special focus on the Tisza region. Furthermore, they present an approach showing how such changes can be avoided with the help of adaptation strategies based on changes in different risk-layers over time.

10.4   Integration

In conclusion, we underline the importance of integrated research that connects the economic, social and ecological systems for sustainable management of the Tisza and all of today's river basins. In many ways the VTT and the government's insurance and compensation programs are interrelated and even mutually dependent, illustrating the importance of integrating environmental, social and economic policy. Strategies for compensating farmers for crop losses due to intentional inundation, or due to permanent loss of their land, are a critical part of any environmentally oriented program, and should be transparent. While insurance is not meant to compensate the "losers" of government programs, it can spread losses from extreme floods and, as these chapters show, even build in solidarity with subsidized premiums. If, as intended by the VTT, the flood system would tolerate some extreme flooding (which ecologists claim is vital for the ecosystems of the region), it is important to spread these losses across a wide community. The intent of the national flood insurance program was to spread these losses across residents throughout Hungary. The failure of the insurance program to provide such a comprehensive safety net greatly complicated implementation of the VTT.

References

Andrásfalvy B (1973) Ancient floodplain and water management at Sarkoz and the surrounding area before the river regulations (in Hungarian). Vízügyi történeti füzetek 6

Balogh P (2002) Basics and method of floodplain management on middle Tisza valley. VATI Kht., Gellerthegy Ut 30–32. Budapest

Bellon T (2004) Living together with nature: farming on the river flats in the valley of the Tisza. Acta Ethnographica Hungarica 49:243–256

Bran J, Borsos B (2009) Conservation and restoration of the globally significant biodiversity of the Tisza river floodplain through integrated flood plain management, Report commissioned by the Hungarian Ministry of Environment and Water, Government of Hungary and United Nations Development Program, UNDP Atlas ID 0046904

Halcrow Water (1999) Flood control development in Hungary: feasibility study. Final report. Halcrow Group Ltd., London

Koncsos L, Balogh E (2007) Flood damage calculation supported by inundation model in the Tisza Valley. In: Koncsos L, Balogh E (eds) Proceedings of the 32nd Congress of the International Association of Hydraulic Engineering and Research. International Association of Hydraulic Engineering and Research. Venice, Italy

Linnerooth-Bayer J, Vári A, Thompson M (2006) Floods and fairness in Hungary. In: Verweij M, Thompson M (eds) Clumsy solutions for a complex world: governance, politics and plural perceptions. Palgrave Macmillan, Basingstoke, New York

Sendzimir J, Flachner Z (2007) Exploiting ecological disturbance. In: Scherr S, McNeely J (eds) Farming with nature: the science and practice of eco-agriculture. Island Press, Washington, DC

Sendzimir J, Pahl-Wostl C, Kneiper C, Flachner Z (2010) Stalled transition in the Upper Tisza river basin: the dynamics of linked action situations. Environ Sci Policy 13(7):604–619

Vári A, Linnerooth-Bayer J, Ferencz Z (2003) Stakeholder views on flood risk management in Hungary's Upper Tisza basin. In: Linnerooth-Bayer J, Amendola A (eds) Special edition on flood risks in Europe. Risk Anal 23:537–627

Werners SE, Warner J, Roth D (2010a) Opponents and supporters of water policy change in the Netherlands and Hungary. W Altern 3(1):26–47

Werners SE, Matczak P, Flachner Z (2010b) Individuals matter: exploring strategies of individuals to change the water policy for the Tisza river in Hungary. Ecol Soc 15(2):24

Chapter 11
Social Indicators of Vulnerability to Floods: An Empirical Case Study in Two Upper Tisza Flood Basins

Anna Vári, Zoltan Ferencz, and Stefan Hochrainer-Stigler

Abstract This chapter aims to develop indicators of social vulnerability related to flood impacts on the regional level. Impacts are seen here as a function of the exposure as well as the vulnerability dimensions. Because key vulnerability factors include several variables that cannot be found in statistical databases, such as preparedness to the hazard, mental coping capacity, social relations, and trust, an approach based on questionnaire surveys instead of only using statistical data from institutions was chosen. The analysis is based on an empirical survey conducted in the Bodrogköz area and in the Bereg region within the Tisza flood basins. We found that while the most important variables influencing impacts were the exposure level and the geographic location, the most important factors of vulnerability were found to be the following: health, education, savings, opportunities of taking loans, trust in the members of the community and in institutions, and perception of preparedness of institutions against floods. Based on the results we give some policy recommendations with regard to increasing the resilience of the exposed communities. These include: increasing public spending on education, strengthening social cohesion, introducing contingency loans so that borrowing is feasible also for the poorer communities and improving flood preparedness by providing relevant information for inhabitants.

Keywords Vulnerability to floods • Empirical survey • Case study • Upper Tisza river basin

A. Vári (✉) • Z. Ferencz
Institute of Sociology, Hungarian Academy of Sciences,
Uri u.49, 1014 Budapest, Hungary
e-mail: anna.vari@socio.mta.hu

S. Hochrainer-Stigler
International Institute for Applied Systems Analysis (IIASA),
Laxenburg, Austria

A. Amendola et al. (eds.), *Integrated Catastrophe Risk Modeling: Supporting Policy Processes*, Advances in Natural and Technological Hazards Research 32,
DOI 10.1007/978-94-007-2226-2_11, © Springer Science+Business Media Dordrecht 2013

11.1 Introduction

In large parts of Europe, extreme weather events, such as heavy precipitation, wind storms and heat waves, are expected to become more frequent and intense in the future due to climate change (Parry et al. 2007; Alcamo et al. 2007). However, climate-related extremes already put a heavy burden on Europeans at different scales, from households, businesses and governments to the European Union. They differentially affect society depending on geography, as well as the economic, social and cultural context of those exposed, including age, health status, education, income, indebtedness, to name but a few factors contributing to vulnerability (Linnerooth-Bayer et al. 2005). Hence, a better understanding of the complex relationships of these factors will also help to decrease vulnerability against extremes more effectively not only for today but also in the future.

The term "vulnerability" is nowadays a concept with multiple and ambiguous meanings, used within a broad range of disciplinary contexts, including geography, anthropology, engineering sciences, ecology and economics. For example, while in the context of climate change, vulnerability is defined as "the degree to which a system is susceptible to, and unable to cope with, adverse effects of climate change, including climate variability and extremes. [. . .] is a function of the character, magnitude and rate of climate change and the variation to which as system, is exposed, its sensitivity and its adaptive capacity" (IPCC 2007: 27). In the disaster community vulnerability is defined as "the characteristics and circumstances of a community, system or asset that make it susceptible to the damaging effects of a hazard" (UNISDR 2003: 12). Hence, in the later terminology vulnerability is independent of its exposure. To make things even vaguer, in the disaster community it is common to use the notion of vulnerability more broadly, and usually vulnerability includes the element's exposure (UNISDR 2003). A more workable definition of vulnerability for this article comes from Turner et al. (2003), which defines vulnerability as the degree to which a system or subsystem is likely to experience harm due to exposure to a hazard, either as a perturbation or stressor. Most importantly in this approach vulnerability incorporates not only exposure but also resilience, now a key concept in vulnerability research, which refers to the capacity of the system to absorb disturbances and reorganize, while undergoing changes to retain essentially the same function, structure, and identity (Walker et al. 2002). Hence, resilience decreases vulnerability.

Still, at this level of complexity it is difficult to carry out any empirical research and focus on some dimensions of vulnerability is necessary. Generally speaking, the different dimensions can be grouped into physical, economic, social and environmental factors as listed below (Kohler et al. 2004):

- Physical: related to the susceptibility to damage of engineering structures such as houses, dams or roads. Also factors such as population growth may be subsumed under this category.

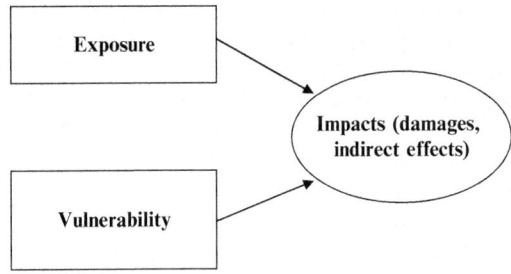

Fig. 11.1 Exposure to hazard, vulnerability and impacts

- Social: defined by the ability to cope with impacts on the individual level as well as referring to the existence and robustness of institutions to deal with and respond to natural disaster.
- Economic: refers to the economic or financial capacity to refinance losses and recover quickly to a previously planned economic activity path. This may relate to private individuals as well as companies and the asset base and arrangements, or to governments that often bear a large share of a country's risk and losses.
- Environmental: a function of factors such as land and water use, biodiversity and stability of ecosystems.

Furthermore, natural disasters may cause a variety of effects which are usually classified into social, economic, and environmental impacts as well as according to whether they are triggered directly by the event or occur over time as indirect effects. In this chapter social and economic vulnerability is looked at only, and exposure is treated as a separate variable, both together with vulnerability leading to damages and indirect effects (Fig. 11.1).

It is a central issue and one of the key goals in the vulnerability research community to find out what factors determine the vulnerability of individuals, communities, organizations and systems, and how vulnerability can be reduced (UNU–EHS 2005). Research suggests that in general, key social and economic dimensions of vulnerability include preparedness for managing the hazard, demography, economic situation, and education and skills, among others (Cutter 2005; Glatron and Beck 2008). The authors believe that in spite of commonalities, there are also substantial differences between main factors of vulnerability in various situations. Economic and social circumstances, institutional background, cultural characteristics and the type of hazard seem to be important determinants of key vulnerability factors.

The purpose of this chapter is to develop regional indicators of social and economic vulnerability to flood damages in the Upper Tisza region. We hypothesize that many key vulnerability factors cannot be found in statistical databases, such as preparedness to the hazard, mental coping capacity, social relations and trust, among others. For this reason we use a standardized questionnaire so that these variables can be incorporated within this study design.

The chapter is organized as follows: The next section introduces the questionnaire, sampling method and first exploratory results. Section 11.3 presents the results separated according to bivariate and multivariate relationships found in the statistical analysis of the data. Finally, Sect. 11.4 ends with a discussion of the results and conclusions.

11.2 Methodology

11.2.1 Sample

A face-to-face questionnaire was administered in two high-risk flood basins (Bodrogköz and Bereg) of the Upper Tisza region,[1] with samples of 400 interviewees in 18 villages in the Bodrogköz area and 300 interviewees in 22 villages in the Bereg region. Data collection was conducted in January 2006 in Bodrogköz, and in August 2006 in Bereg. The interviewees were chosen randomly from the population by the demographical quota.[2] This quota ensured representativeness of the population in the sample with respect to gender of the respondents, their age (approximately half of the respondents consisted of adults below 29 years of age and above 60 years of age), and education (most respondents had less than 8 classes of primary school, with Bereg showing a larger amount compared to Bodrogköz).

11.2.2 Method

The purpose of the questionnaire was to obtain information from the respondents on their exposure, vulnerability and impacts from previous floods:

• *Exposure*: The water management authorities classify settlements according to their flood exposure; however, due to differences in elevation there is differential exposure even within one settlement. For this reason, we chose to rely instead on the respondents' self classification of their exposure as part of the questionnaire;
• *Vulnerability*: We initially hypothesized that flood vulnerability is related both to individual and community preparedness and to social and economic characteristics, such as health, education, economic activity, income, savings,

---

[1] Data collection was supported by the following organisations: United Nations University and the Research Institute for Soil Science and Agro-chemistry of the HAS (Bodrogköz); UNDP GEF, Directorate of the Hortobágy National Park and the Ministry of Environment (Bereg). Data processing was financed by the Department of Mathematics and Information Technology of Corvinus University.

[2] The data were collected from the census database (2001) of the Hungarian National Statistical Office.

Table 11.1 Exposure, vulnerability and impact sub-dimensions

I. Exposure		Exposure of the respondent's settlement to floods
		Exposure of the respondent's home to floods
		Personal experience concerning floods
II. Vulnerability	Preparedness	Preparedness of the respondent (and his/her family) for floods
		Preparedness of different institutions (government, local government, water authority, water associations) for floods
	Physical and mental health	Respondent's health status
		Respondent's lasting health damage or impairment
		Respondent's (mental) capacity of coping with problems
	Qualification	Respondent's educational level
	Economics	Respondent's economic activity and income
		Respondent's savings
		Respondent's opportunities for borrowing
	Social capital	Trust in members of the community and in institutions
		Respondent's social relations and isolation
		Civic activity of respondent
III. Impacts of floods		Respondent's (and family's) damages and disadvantages caused by recent floods
		Lasting effects of recent floods

and social capital.[3] As a basis for the questions, we made use of vulnerability indicators found to be relevant in the international literature (for example UNU–EHS 2005), as well as in the findings of our earlier research (Vári and Ferencz 2006);

- *Impacts*: Only a very small number of people have lost their lives in floods in Hungary during the past decades, and damages have been primarily of economic and social nature. Therefore we focused questions on exploring such impacts.

Table 11.1 lists the components of exposure, vulnerability and impacts that formed the basis of the survey questions.

11.3 Summary of Questionnaire Responses

In the following, we summarize the results of the questionnaire responses in Bereg and Bodrogköz before turning in the next section to examining the relationships among exposure, vulnerability and impacts. We present the results of bivariate analyses in which we test the significance of correlations using Chi-square tests and

---

[3] The concept of social capital includes trust, intra-community relations, and the strength of civil society and certain aspects of governance (see Putnam 1993; Fukuyama 1996).

Table 11.2 Responses to
question asking how prepared
were respondent,
respondent's family and
relevant institutions for floods
in the past (average of a five-
grade scale)

	Bereg	Bodrogköz
You and your family	2.07	2.31
The central government	2.42	2.11
The local government	2.49	2.32
The water management authority	2.59	2.51
The water associations	2.56	2.35

ANOVA model approaches.[4] The purpose is to give a comparison between the two
selected flood hazard prone areas and to detect differences with regards to the
vulnerability dimensions.

11.3.1 Exposure

In Bereg, the overwhelming majority of respondents regard their settlement as
being either strongly or weakly exposed, and less than one tenth of the respondents
believe that there is no danger of floods. Two thirds of Bodrogköz respondents
regard their settlement as strongly or weakly exposed to floods, whereas one third
hold that there is no such danger. Those who regard their home (weakly or strongly)
exposed made up about 94% of those living in the exposed settlements of Bereg,
and about 72% of those living in exposed settlements in Bodrogköz. A more
detailed analysis showed that active earners and diploma-holders are overrepresented among those not exposed (i.e. those considering that either their settlement
or their home are not exposed) in the Bodrogköz area, whereas the unemployed and
people with primary education were over proportionally represented among the
exposed. In the Bereg region there was no significant relationship between exposure
and socio-economic variables. Inquiring if the respondents *had already experienced flooding*, 90% of the Bereg respondents reported living through a flood (89%
experienced the 2001 flood in this region) and 33% had experienced multiple
floods. In the Bodrogköz region 32% of the respondents had experienced flooding
and 20% multiple floods. Exposed respondents were overrepresented among those
who had experienced at least one flood, indicating a (significant) correlation
between having experienced floods and perceiving a higher exposure.

11.3.2 Preparedness

Tables 11.2 and 11.3 present the assessment of past and future flood preparedness
among people who had experienced floods in both regions. On a five-point scale,

---

[4] In the present chapter those interrelationships are mentioned from which significant relationships
among variables can be shown, in other words we may state on the 95% confidence level that the
variables are not independent of each other.

Table 11.3 Responses to question asking how prepared are respondent, respondent's family and relevant institutions for future floods (average of a 5-grade scale)

	Bereg	Bodrogköz
You and your family	3.0	3.09
The central government	3.16	2.91
The local government	3.24	3.16
The water management authority	3.37	3.26
The water associations	3.32	3.14

the average assessment was between 2.07 and 3.37. Assessments of past preparedness were lower in every category than for future preparedness, and Bereg was considered better prepared than Bodrogköz. Differences, however, between values of future and past preparedness were very similar in both regions, around 0.7. As far as institutions are concerned, people regard water management authorities as the most prepared in both regions, followed by water associations and local governments. The lowest scores were given to the central government, which reflects a general disappointment with central government agencies. In the Bodrogköz region the respondents assessed their own preparedness more positively than that of the central government, whereas people of Bereg regarded their own preparedness as less positive.

Interestingly exposed people considered themselves better prepared than those not exposed in both regions.[5]

11.3.3   Physical and Mental Health

The respondents were asked to *evaluate their own health status* on a five-grade scale. The average assessment was 3.32 in Bereg and 3.34 in Bodrogköz. In Bereg and Bodrogköz, there were larger proportions of women, pensioners, people above 50, those with primary education, and people having low (household) income, who assessed their health status as poorer. Alternatively, men, people between the ages of 18 and 39 (18 and 49 in Bodrogköz), active earners, those who had completed their secondary studies and those who had a medium or high income were over-represented among those considering themselves to have good health status. Women reported a significantly worse health status than men. Fifty per cent of men, whereas only 38% of women regarded themselves as in good health in the Bodrogköz region, while these proportions were 55 and 41% respectively in Bereg.

The respondents were also asked whether they *had lasting health damage or impairment*. From this question the population of Bodrogköz seems to be somewhat healthier: 28.9% reported having permanent damage to their health, as contrasted to Bereg, where this proportion was 33%. In Bereg, pensioners, people with primary education and those of the lowest income indicated permanent health damage above

---

[5] Opinions assessing the current situation in the Bereg region are an exception where the difference is within the margin of error.

Table 11.4 Employment status of interviewees (%)	Bereg	Bodrogköz
Active earner	26.3	23.4
Pensioner	42.0	39.7
Unemployed	15.7	15.4
Other inactive	16.0	21.5

the average. In the Bodrogköz region it was mostly pensioners, those of primary education, people above 50 and of low-medium income who indicated having lasting health damage. In Bereg as well as in Bodrogköz the relationship between health status and health damage was significantly correlated.

Another potential factor of vulnerability is the *capacity to cope with problems,* which we explored with a question that elicited coping strategies of those who experienced a flood. In both regions, a typical response for coping was to try to analyze and understand the situation, especially among the younger people (40–49) and those considering themselves as less exposed. In Bereg a typical coping strategy was to take a positive attitude or interpretation of the problems faced. In Bodrogköz, a frequent response was coping through positive personal change or "emerging as a different person", combined with creative activity. Taking sedatives and medicines, as well as self-destructive activities, were characteristic only to a small extent, but more in Bodrogköz than in Bereg and more by those considering themselves to be less healthy.

11.3.4   Education

The proportion of respondents having completed not more than 8 classes of primary school was 59% in Bereg and 47% in the Bodrogköz region. In Bereg 21% and in Bodrogköz 31% of the respondents held certificates from a vocational secondary school. The proportion of those who had passed their grammar secondary final certificate was 14 and 18%, respectively, whereas the proportion of those who had university degrees was 5 and 4%, respectively.

11.3.5   Household Economic Data

As far as *employment status* is concerned the survey responses are reported in Table 11.4.

The proportion of active earners is lower in both regions than the national average (58%), and unemployment is more than double the national average (7.5%). Responses to questions on *household incomes* follow a similar pattern, except the proportion of medium incomes (HUF 91–120,000) and high income (above HUF 121,000) are somewhat higher in Bereg (28 and 30%) compared to 25 and 26% in Bodrogköz.

Table 11.5 Reported forms of savings (%)

	Bereg	Bodrogköz
In real estate	2.3	7.9
Other assets	11.0	10.8
At home in cash	19.7	12.8
In savings books and savings accounts	22.0	20.1
In life-, pension – or health insurance funds	16.7	14.1

Table 11.6 Reported forms of borrowing options (small and large)

	Small amount		Large amount	
Borrowed from	Bereg	Bodr.	Bereg	Bodr.
Immediate family members	74.7	62.6	13.0	10.3
Relative living in the same settlement	33.3	32.4	5.3	6.9
Distant relative	13.3	9.0	1.3	2.2
Acquaintance, neighbor, or associate at work	22.3	16.3	2.7	1.8
Bank or credit institution	41.0	20.6	27.3	21.4

Savings can enable households to recover from floods and thus represent an important factor reducing vulnerability and building coping capacity. Table 11.5 shows the types of reported savings.

Not surprising, those with primary education, the unemployed and other inactive persons are overrepresented among those not having savings, which in Bereg was 66% of the population and in Bodrogköz 57%. In the Bodrogköz region, people living in exposed regions mentioned real estate and other assets as forms of savings in larger proportions than those in Bereg, whereas cash at home and savings accounts were mostly characteristic of pensioners. It is active earners, people with grammar, secondary school, and university degrees as well as people between 30 and 39 years of age, who invest in insurance.

As in the case with savings also borrowing capabilities can enable households to recover from floods and thus also represent an important factor. Table 11.6 reports findings on who respondents borrowed from, dependent on the amount.

In Bereg the possibilities of taking loans were assessed as better in every category than in the Bodrogköz region. Generally speaking the possibility of getting loans from close relatives and acquaintances occurred in greater proportion in the case of smaller sums only, whereas distant relatives did not figure significantly either in the case of smaller or of significant sums. It was active earners, those with grammar secondary education and university degree who mentioned the various possibilities of taking loans above the average. In Bereg active earners and people of at least secondary education were those who had outstanding proportions among those capable of receiving smaller loans. In addition to those groups it was mostly people of medium- and high household income and those between 40 and 49 years of age who were capable of getting bigger loans.

Table 11.7 Trust in
members of the community
(averages of a 100-grade
scale)

Trust in	Bereg	Bodrogköz
People living in the neighborhood	36	39
More distant acquaintances	40	40
People of workplace	45	45

11.3.6 Social Capital

Trust can be an important indicator of social capital (Newton 2001). We measured
(i) trust in members of the close community (neighbors, acquaintances, associates
at work) and (ii) trust in public institutions. A strikingly low level of trust was found
in community members and public institutions in both regions (Table 11.7).

In the Bereg region, active earners were overrepresented among those who
trusted members of the community, whereas pensioners, the unemployed and
those of the lowest income were over-represented among the mistrustful. In the
Bodrogköz region trust-related responses do not offer as uniform of a picture as in
Bereg. Active earners trusted most their neighbors; pensioners trusted most their
more distant acquaintances, whereas active earners, men, those of vocational
secondary education and people of the highest income had greatest trust in their
associates at work. The unemployed, other inactive people, as well as low-medium
income people were more mistrustful of their neighbors. People between 18 and
29 years of age as well as the unemployed were mistrustful of more distant
acquaintances. Women, people of primary education as well as low- and
medium-income were less trustful of their associates at work.

The most trusted institutions in both regions were the schools, police, water
management authorities and water associations. The credibility of the national
government was regarded the lowest in both areas. Considering the socio-
demographic variables, the younger age groups, the less qualified and those of
lower incomes, as well as inactive people, reported less trust in public institutions
than the average. The main difference between the two regions is that opinions
related to credibility are divided by age and income in the Bodrogköz region,
whereas they are divided more by school education in the Bereg region. Economic
activity is a significant factor in both regions.

We explored the *social relations* of respondents by asking *how many family
members and relatives lived in the given settlement or region*. The average number
of family members and relatives living in the same settlement was 22 in Bereg and
21 in Bodrogköz, i.e., large families are still typical in both regions.

We measured *social isolation* by asking how much the respondent agreed to the
following statement: "I frequently feel myself lonely." In Bereg 26% of
respondents reported that this statement was fully or partly true, whereas this
proportion was 24% in the Bodrogköz region. These figures are surprisingly high,
considering the traditionally strong ties within extended families and among
neighbors in small villages. In Bereg, women and pensioners were in the greatest
proportion among those who feel entirely or partially lonely, whereas in the
Bodrogköz region they were joined by those with low incomes and only primary

Table 11.8 Have you ever tried to contact the local government about an issue that affected you? (%)

	Bereg	Bodrogköz
Yes, once	9.4	9.2
Yes, several times	13.7	17.7
No	76.9	73.1

Table 11.9 Types of flood damages suffered since 1998 among those who experienced floods (%)

Type of damage	Bereg	Bodrogköz
Settlement of residence	81	42
Residential home or flat	77	38
Respondent (and family) evacuated	74	9
Relatives	71	29
Agricultural buildings (e.g., pen, stable)	57	22
Furnishings	49	19
Crops, arable land, vineyard, orchard	39	37
Stock and harvested grain	28	7
Savings reduced	23	16
Absence from work, loss of salary and income	13	4
Illness generated or renewed	6	7

education. The extent of loneliness shows negative correlation with the number of relatives in the settlement and region in Bereg as well as in the Bodrogköz region.

We measured the *civic activities* of respondents by the question whether the *interviewee had contacted the local government about* an issue affecting him or her. The results are given in Table 11.8.

In the Bodrogköz region it was people in the 50–59 year age group and those of low to medium income who were over represented among the most active. In Bereg most frequently those between 30 and 39 years of age, diploma-holders and the unemployed had contacted the local government.

Our results indicate low-trust and fragmented communities, with highly limited civic activities. It is not different from the overall Hungarian picture, where the socio-economic transition of 1990 has deepened social inequalities and broken up former solidarities without creating a strong new civil society (Utasi 2006).

11.3.7 The Impacts of Floods

We measured the negative impacts of floods (losses, damages, indirect effects), their gravity and duration by several questions addressed to those who had experienced floods. From Table 11.9 it can be seen that there was a significant difference between the two regions with respect to flood damages. In Bereg the overwhelming majority of the population suffered some kind of damage, whereas that proportion was around one third in the Bodrogköz region. As far as material damages are concerned, in both regions residential property, agricultural buildings, furnishings of the home, as well as crops, arable land, vineyards and orchards suffered damages most frequently.

Table 11.10 Assessment of the durability of flood impacts among those who experienced floods (%)

Duration	Bereg	Bodrogköz
3 months	15	27
6 months	16	35
1 year	47	20
Still can be felt	20	16
"There was no flood"	2	2

Studying the relationships between damages, exposure, and socio-economic variables revealed some insights related to vulnerability. In Bereg, those who suffered the most damages to their homes and agricultural buildings were not only those most exposed, but there was a correlation with respondents reporting low trust in local institutions, limited savings, and limited access to even small sums of loan. In the Bodrogköz region, the correlations were similar with the exception that those most affected also considered themselves to be less prepared. In both regions, floods appeared to impose more losses on those in poor health. The largest difference between the two regions was the number of those experiencing evacuations – 9% in the Bodrogköz region compared to 75% in Bereg. Those evacuated appeared to be disproportionally in the group who were mistrustful of members of the community and public institutions, had no savings and could not obtain small loans.

Another question, reported in Table 11.10, asked *about the duration of the physical impact of the floods*. The perception of duration appears to be shorter in Bodrogköz, although it is striking that around one-fifth of those experiencing floods in the past feel that the impacts have continued to the present.

In the Bereg region this response was related to trust. Those who perceived the effects of floods for a shorter time were those who trusted their neighbors, acquaintances and associates at work, and felt most public institutions were credible.

According to the above analysis the two investigated flood basins significantly differ in terms of exposure, i.e., in Bereg a much higher proportion of homes is exposed to floods than in Bodrogköz, and similarly, a much higher proportion of the inhabitants have already experienced flooding and suffered damages. In terms of socio-economic characteristics differences between the two areas are smaller. Concerning the level of health, education, and savings the situation is somewhat better in Bodrogköz than in Bereg, whereas the ratio of active earners, the magnitude of household incomes and the opportunities for taking loans are somewhat more favorable in Bereg. More importantly, however, both regions are strongly handicapped if compared with the national average, especially in terms of qualification and economic activity.

11.4 Vulnerability Indicators

After the detailed presentation and comparison of the vulnerability and exposure variables for the two regions, we now turn to the question of what variables or sets of variables can explain best the responses on impact. As shown in Fig. 11.1 we will

Table 11.11   Selected impact (damages and indirect effects) variables

Variables	Abbreviation
Damages in residential property and/or in its contents	D1
Agricultural damages (damages to agricultural buildings, crops, harvest stock)	D2
Loss of income	D3
Evacuation and/or health damage	D4
Duration of impacts	D5

Table 11.12 Vulnerability variables selected on the basis of principle component and correlation analysis

Variables	Abbreviation
Health status	V1
School education	V2
Economic activity	V3
Household income	V4
Having any form of savings	V5
Possibility of getting a small loan	V6
Possibility of getting a large loan	V7
Trust in members of the community	V8
Assessment of the credibility of institutions	V9
Assessment of past preparedness	V10
Assessment of future preparedness	V11

treat impacts as a function of exposure and vulnerability. This assumption seems to be valid as exploratory bivariate correlation analyses have shown that most impacts are related to perceived flood exposure, *and* to most of the hypothesized vulnerability variables, while keeping exposure constant. To identify factors, i.e. sets of variables representing a latent construct, not measurable with a single variable, we first applied principle component analysis[6] of impacts by creating these variables first (see Table 11.11) and afterwards looked at the vulnerability and exposure variables which show significant correlation:

The exposure variable was chosen to be the respondents' exposure (a combination of the settlement' exposure and the home's exposure variables, called E1). The following vulnerability variables were selected based on (1) significant correlation to damages, and (2) those which carry the largest information content within the given group of variables. Table 11.12 shows the results. Some interrelationships and important differences between the two regions were identified among the above variables. In Bereg significant relations exist among the V1–V7 variables. In Bodrogköz significant relations were found among the V3–V8 variables, and V1 is also correlated with variables V3, V4, V5 and V7. In Bereg the V8–V11 indices of trust and preparedness show correlation with each other, whereas in Bodrogköz they show close correlation rather with members of the V1–V8 group. There are significant connections between respondent's exposure (E1) and certain indicators

---

[6] Some variables were transformed in advance, for instance we have transformed variables measured on scales of four and five grades into a 100-grade scale.

Table 11.13 Latent factors, number of variables and Cronbachs alpha[a]

Index (Abbreviation)	Number of variables	Cronbachs Alpha
Impact Factor (IdF)	11	0.842
Preparedness Institutions Factor (VprepF)	4	0.946
Savings Factor (VsavF)	5	0.955
Borrowing Factor (VborF)	10	0.736

[a]For example, the impact factor (IdF) is now a continuous variable which is basically, for each observation, the sum of the responses to the 11 impact questions coded as 0 or 1 (no or yes). Hence, the higher the number of IdF the higher the (negative) impact The other factors were formed in a similar way: the Preparedness Institutions Factor (VprepF) is the sum of the responses to the 4 questions associated with the past preparedness of the various institutions, the Savings Factor (VsavF) is the sum of 5 responses on savings, and the Borrowing Factor (VborF) is the sum of 10 responses on the borrowing possibilities. Also, instead of the dichotomous exposure variable E1 we used the Exposure variable based on the respondent's self-evaluation of exposure to floods, which had three possible values, including 'strong', 'weak' and 'no' exposure

of vulnerability (V1, V4 in Bereg, V2, V3 and V6 in Bodrogköz). The socio-economic status of those exposed is somewhat worse; there are greater proportions of less healthy, less qualified and less active people among them. This suggests that socially disadvantaged groups live in larger proportions in high risk areas. Respondent's exposure (E1) shows significant correlation with most indicators standing for impacts (D1–D5) in both regions. All the vulnerability indicators (V1–V11) show significant correlation to the variables indicating impacts (D1–D5) (even if the effects of exposure are screened), at least in one region.

The above analysis indicates that there are strong relations among various vulnerability indicators, as well as between variables of exposure, vulnerability and impacts. In order to further analyze these relationships, latent factors based on the results above are constructed. However, we re-assessed the reliability of the scales too. Afterwards, we determined the set of variables for each of the factors by choosing only those variables from each set that returned the highest reliabilities (using Cronbachs Alpha). The factors that have been built with this procedure are listed in Table 11.13.

Using the new factors, as well as the other vulnerabilities explained in detail in the previous section, we proceeded with multivariate tests and analyses. As Fig. 11.2 indicates, it is evident that the "Area" variable (Bereg or Bodrogköz), as well as the "Exposure" variable have a dominant role for the impact factor IdF (Fig. 11.2).

The box plot shows that IdF for each exposure sub-group is higher for the Bereg area.[7] Furthermore, one can see that for decreasing exposure there is a decrease in the IdF irrespective of the Area variable. Differences between the IdF and the Area variable, as well as the Exposure variable, are highly significant (a non-parametric

[7] Which can be seen, for example, by the thick black line in each box plot which represents the median.

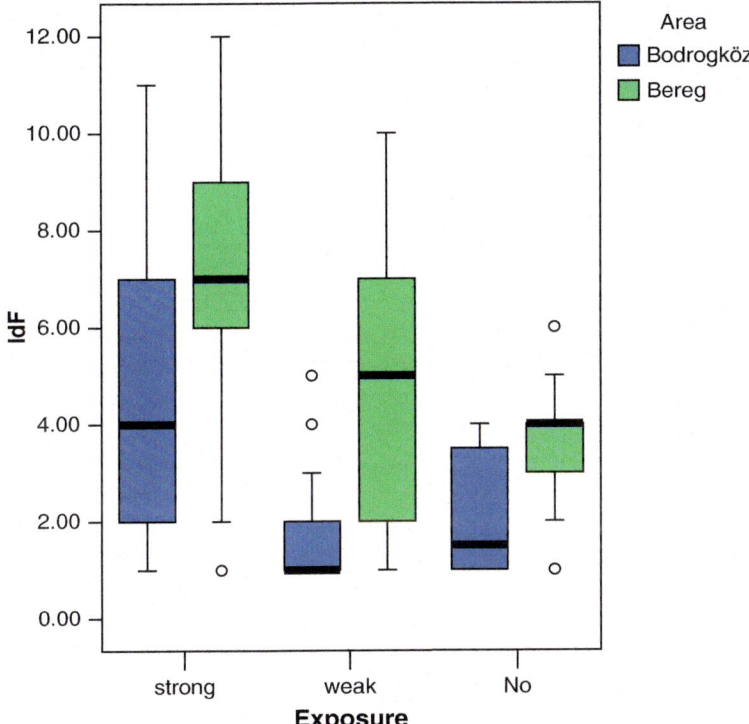

Fig. 11.2 *Box* plots of the impact factor separated according to area and exposure

Mann Whitney-U test was used). However, this is not the case for the interactions between the two variables and IdF, i.e. Area and Exposure together does not show significantly different IdF.

As a next step, to incorporate interactions between the vulnerability and exposure variables as well as the latent factors, a general linear model approach was used.[8] In more detail, a general linear model with two factors (Exposure and Area), as well as the corresponding vulnerability covariates, was created and tested. Interrelationships up to the third level were also enabled. The model was significant with an R-square of 0.699. Significant variables included the exposure and area variables, trust, education, borrowing capacity, savings, health and perception of (past) institutional preparedness. In a next step the sample was analyzed by means of dummy variables again using a general regression model, but now without the factors, but keeping interactions possible up to the second level. For example, we looked at each Exposure and Area sub-group and performed a regression analysis.

---

[8] Here, combinations of factors (dichotomous variables) and covariates (continuous variables or factors) can be studied in more detail. Usually, continuous independent variables are called covariates an dichotomous independent variables are called factors in general linear models. Hence, we use these terms in the following.

Table 11.14 Vulnerability variables derived from regression analysis

Variables	Abbreviation
Health status	V1
School education	V2
Savings Factor	VsavF
Borrowing Factor	VborF
Trust in members of the community	V8
Social relations	V12
Preparedness Institutions Factor	VprepF

For Bodrogköz (medium exposure) no significant variables were found. Reasons for that could be the small number of observations, as well as a small spread of the IdF variable. For Bereg (high exposure) significant variables included education, savings level, borrowing capacity, trust, social relations (i.e., number of family members in the region), and civic activity.

As regards the relationships between vulnerability variables and impacts, savings and borrowing abilities (and both together) are important, e.g. the higher the capacities, the lower the impacts, however, correlations are low. Not surprisingly, perception of good self preparedness in the past correlates with lower impacts, and to the contrary, bad perception of the preparedness of the responsible institutions correlates with high impacts. Also, with higher social relationships within the community, impacts decrease. Alternatively, stronger civic activity shows higher impacts, which could be explained in the sense that those who suffer large losses have more motivation to complain to the authorities, which would then mean that civic activity should be regarded as an exposure variable. Vulnerability indicators drawn from the above analysis are summarized in Table 11.14.

The importance of the variables differ dependent on exposure level. Especially health status and education are important vulnerability indicators for middle exposed households, while for highly exposed households, savings, borrowing, trust, and social relations are more important as indicators for vulnerability. Trust and perception of preparedness of institutions are overall indicators of vulnerability (but with lower correlations).

11.5   Conclusions and Recommendations

The primary aim of the chapter was to determine the major socio-economic factors of flood vulnerability in regions highly exposed to floods. As it was expected, the most important single variable determining impacts was the level of exposure and geographical location. Most important indicators of social vulnerability proved to be the following: health, education, savings, opportunities of taking loans, trust in the members of the community and institutions, social relations, and perception of preparedness of institutions against flood events. Remarkably, the majority of indicators are related to human and social capital, as well as institutional capacities. Economic variables, including income and employment appear less significant, which may partially be the result of the low reliability of such data.

We found that the situation of the population of the Upper Tisza regions is rather diverse regarding vulnerability. Only 40–50% of the population assesses their health status as being good; only 40–55% have completed more than primary education; only 35–45% have savings; and less than 35% would have access to large loans. Trust is rather low and people assessed their flood preparedness as slightly higher than mediocre. On the basis of the survey, it is possible to identify the most vulnerable groups that are in a disadvantageous position, due to their health and education status, as well as economic strength and social relations. Hence, these indicators seem to be valid for determining the social vulnerability due to floods.

This research goes beyond the study of the vulnerability of the regions in question. Based on the indicators identified and the questionnaire created for their measurement it may be expedient to assess the vulnerability of populations in other high flood risk areas of Hungary and to identify the particularly vulnerable groups. From a policy perspective, it seems worthwhile to further identify options for reducing the level of exposure, either by structural or non-structural mitigation measures. In addition, there are various opportunities to increase the resilience of exposed communities. For example, increasing public spending on education would increase the resilience of households in the future. Strengthening social cohesion would most likely be an effective intervention. From a disaster risk financing strategy, limited options remain for the government to directly help people at the household level. However, there are large opportunities to help the population help themselves in the future, for example, by introducing contingency loans so that borrowing is also feasible for poorer communities and by establishing public/ private insurance arrangements that are both feasible and attractive for property owners. Creating incentives to increase informal strategies to lessen the short term (and therefore also the long term) consequences of the disaster event, such as providing information on what should be done in case of floods (e.g. safe meeting places for inhabitants, as well as for volunteers) and providing timely information on where to apply for financial support would also increase the resilience of exposed communities.

References

Alcamo J, Moreno JM, Nováky B, Bindi M, Corobov R, Devoy RJN et al (2007) Europe. Climate change 2007: impacts, adaptation and vulnerability. Contribution of working group II to the fourth assessment report of the intergovernmental panel on climate change. In: Parry ML, Canziani OF, Palutikof JP, van der Linden PJ, Hanson CE (eds) Climate change 2007: impacts, adaptation and vulnerability. Cambridge University Press, Cambridge, pp 541–580

Cutter SL (2005) The geography of social vulnerability: race, class, and catastrophe, in Social science research council, Understanding Katrina: perspectives from the social sciences. http:// understandingkatrina.ssrc.org/Cutter/. Accessed 01 Jan 2010

Fukuyama F (1996) Trust: the social virtues and the creation of prosperity. The Free Press, New York

Glatron S, Beck E (2008) Evaluation of socio-spatial vulnerability of city dwellers and analysis of risk perception: industrial and seismic risks in Mulhouse. Nat Hazards Earth Syst Sci 8:1029–1040

IPCC (2007) Fourth assessment report: climate change (AR4)

Kohler A, Jülich S, Bloemertz L (2004) Guidelines. Risk analysis – a basis for disaster risk management. Deutsche Gesellschaft für Technische Zusammenarbeit (GTZ) GmbH, Eschborn

Linnerooth-Bayer J, Mechler R, Pflug G (2005) Refocusing disaster aid. Science 309 (5737):1044–1046

Newton K (2001) Trust, social capital, civic society and democracy. Int Pol Sci Rev 22 (2):201–214

Parry ML, Canziani OF, Palutikof J, van der Linden P, Hanson C (eds) (2007) Climate change 2007: impacts, adaptation and vulnerability. Contribution of working group II to the fourth assessment report of the intergovernmental panel on climate change. Cambridge University Press, Cambridge

Putnam RD (1993) Making democracy work: civic traditions in modern Italy. Princeton University Press, Princeton

Turner MG, Collins S, Lugo A et al (2003) Long-term ecological research on disturbance and ecological response. Bioscience 53:46–56

UNISDR (2003) Terminology on disaster risk reduction. http://www.adrc.asia/publications/terminology/top.htm. Accessed 01 Jan 2010

UNU-EHS United Nations University Institute for Environmental and Human Security (2005) Vulnerability assessment in the context of disaster-risk, a conceptual and methodological review. United Nations University, Bonn

Utasi A (ed) (2006) Resources of subjective quality-of-life: security and relationships. HAS Institute of Political Sciences, Budapest (in Hungarian)

Vári A, Ferencz Z (2006) Flood research from a social perspective: the case of the Tisza river in Hungary. In: Tchigurinskaia I, Ni Ni Thein K, Hubert P (eds) Frontiers in flood research, IAHS publication 305. IAHS Press, Wallingford, pp 155–172

Walker B, Carpenter S, Anderes J, Abel N, Cumming G, Jansen M et al (2002) Resilience management in social-ecological systems: a working hypothesis for a participatory A. Conserv Ecol 6(1):14

Chapter 12
Designing a Flood Management and Insurance System in Hungary: A Model-Based Stakeholder Approach

Joanne Linnerooth-Bayer, Anna Vári, and Lisa Brouwers

Abstract This chapter describes how an integrated catastrophe model aided a stakeholder policy process focusing on the design of the Hungarian flood insurance system. The process incorporated views on flood insurance held by the public, local authorities, government ministries and private insurers. It was based on extensive interviews, a public survey administered to 400 persons in the risk and non-risk communities and a stakeholder workshop. Stakeholder participation was aided by a catastrophe model that could demonstrate the distribution of future flood losses among the victims, the government and the insurers depending on the design of the insurance pool. The Hungarian stakeholders reached consensus on the design of the national insurance system with all its implications for loss reduction and burden sharing. This pilot study illustrates the use of information technology in a partici-patory, stakeholder setting, and as such is of interest to all policy makers seeking social consensus for disaster risk management policies.

Keywords Stakeholder processes • Insurance • Catastrophe modeling • Tisza • Flood risk management

J. Linnerooth-Bayer (✉)
Risk, Policy and Vulnerability (RPV) Program, International Institute for Applied Systems Analysis (IIASA), Schlossplatz 1, A-2361 Laxenburg, Austria
e-mail: bayer@iiasa.ac.at

A. Vári
Institute of Sociology, Hungarian Academy of Sciences, Budapest, Hungary

L. Brouwers
Royal Institute of Technology, Stockholm, Sweden

A. Amendola et al. (eds.), *Integrated Catastrophe Risk Modeling: Supporting Policy Processes*, Advances in Natural and Technological Hazards Research 32, DOI 10.1007/978-94-007-2226-2_12, © Springer Science+Business Media Dordrecht 2013

12.1 Introduction

Over 400 communities and 1.2 million people along the River Tisza are at significant risk from flooding. Since the eighteenth century, the response to flooding has been mainly hydro-engineering operations that have massively reconfigured the river basin. The Tisza floodplain (approximately 16,000 km^2) is now protected by nearly 3,000 km of flood embankments, which, however, remain inadequate to protect against increasing frequency and discharges of floods (Bran and Borsos 2009). Ecologists point out that they threaten what remains of freshwater wetlands that provide valuable eco-services and biodiversity that are essential if the region switches to a more diversified agricultural and livelihood production. As discussed in the Introduction to this section, many argue that a switch from the *river defence paradigm* to a *working landscape paradigm* is necessary given the increasing economic, environmental and social impoverishment of the region (Sendzimir et al. 2010).

The New Vásárhelyi Plan (Hajós 2002) first enacted in 2003 (Government Decision No. 1107/2003-X1.5) would make this *paradigm* shift by changing emphasis from protective levees to the construction of reservoirs and the promotion of biodiversity and other landscape improvement programs. This plan, however, has not been implemented as fully as many would like, which has been attributed primarily to cost overruns in the reservoir construction process and a lack of political support for the landscape improvement (Borsos B, 2011, UNDP Technical Advisor, personal communication, 10 February, Budapest). Slowed implementation can also be partly attributed to the plan's failure to deal satisfactorily with "winner and loser" issues. One issue is compensation to farmers who face reduced livelihoods due to a pull back in river defences, and to those who will be relocated due to the construction of reservoirs.

A half decade before the popular movements addressing implementation of the New Vásárhelyi Plan political pressure had mounted to move away from building embankments and river defences, but for different reasons and by different actors. As Hungary prepared to enter the European Union in 2004, the government recognized the difficulty of continuing its tradition of taking almost full responsibility for flood risk management. A continuing dependence on river defences would require investing huge sums in expanding and maintaining embankments, as well as compensating victims for flood losses (e.g., by reconstructing damaged homes). EU membership demanded fiscal austerity, and the financial authorities welcomed the idea of more private responsibility for the reduction and response to flood disasters.

The topic of this chapter, and an issue that still plagues river basin management in the Upper Tisza region, is how to distribute the gains and losses from national and local policies aimed at reducing or coping with flooding. While the research reported here was conceived and carried out before recent attempts to shift the Tisza flood risk management to more environmentally oriented pathways, the methods and results are highly relevant to current events. Today, this issue manifests itself in the form of compensating farmers for giving up their land for reservoirs and for switching practices to cope with a changing ecological environment and landscape. A decade earlier the issue was how to design a national flood insurance system that would fairly compensate victims of floods in this region and at the same time provide incentives for reducing flood risk.

This chapter shows how an integrated catastrophe model aided a stakeholder policy process focusing on the design of a Hungarian flood insurance system. The 3-year process (2001–2004), consisting of interviews, a public survey and a stakeholder workshop, has been described in detail elsewhere (Linnerooth-Bayer et al. 2006; Vári et al. 2003).[1] We focus on how the catastrophe model informed this process. The dynamic and spatially explicit flood simulation model, including its design, user interface and use, is described in more detail in Brouwers and Riabacke (Chap. 13, in this volume). We briefly summarize the background and contending views on flood risk management in Sect. 12.2. Section 12.3 shows how the stakeholder interviews led to three different options for the design of an insurance and compensation program. In Sect. 12.4 the discussion turns to describing the flood catastrophe model that informed the second round of interviews and the stakeholder workshop. In Sect. 12.5 we discuss the participatory workshop, where the options were examined aided by model, followed by a discussion of the stakeholder consensus in Sect. 12.6. We conclude in the final section by discussing the unique features of this model-based participatory process.

12.2   Stakeholder Views

Many western countries, including the US, France and Norway, have allocated responsibility for the economic losses from floods and other hazards by legislating national insurance systems. Hungary, too, has recently instituted a public-private insurance program (partly building on the results of the stakeholder process reported in this chapter), yet penetration remains low and issues of loss sharing are still on the public policy agenda. How much should people living in non-risk areas and taxpayers contribute to preventing losses and compensating victims in vulnerable communities, and to what extent should those living or locating in high-risk areas bear the burden? This and other questions underline one of the more controversial issues in Hungary – the respective roles of the government and the private market in preventing and pooling disaster losses.

For the purpose of eliciting stakeholder views on flood risk management and especially on how to distribute flood losses, face-to-face, open-ended interviews were carried out with 24 active stakeholders, including persons representing central, regional and local government agencies, farmers and entrepreneurs, NGO

---

[1] The research is based on a project funded by the Swedish Research Council for Environment, Agricultural Sciences and Spatial Planning, and carried out by IIASA, Stockholm University and the Hungarian Academy of Sciences Institute of Sociology.

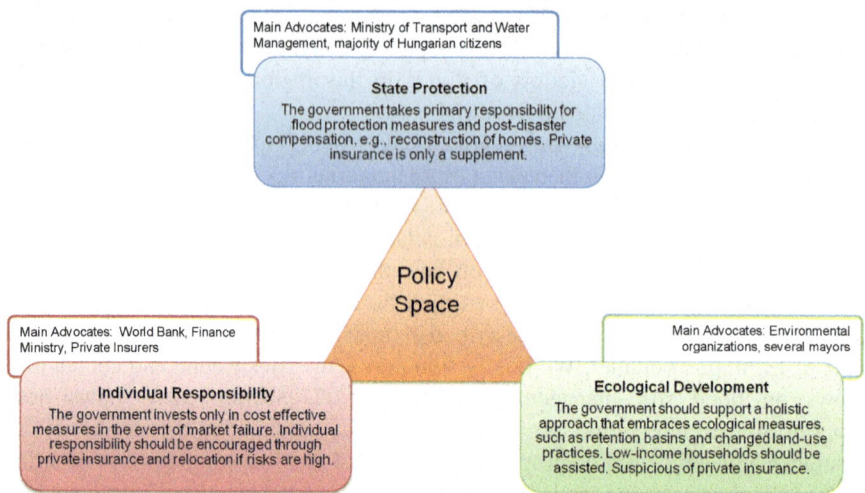

Fig. 12.1 The contested policy terrain (Source: Adapted from Linnerooth-Bayer et al. 2006)

activists and insurance companies. Building on the views revealed by the stake-holder interviews, a survey instrument was administered to 400 persons in Hungary at four separate high-risk and low-risk locations in both rural and urban areas. The interviews and survey results are reported in Vári et al. (2003).

While the interviews and survey predated activities leading to the New Vásárhelyi Plan, in many ways they foretold the emergence of the "shadow network" of environmentalists, intellectuals and others who now advocate a changed management paradigm (Sendzimir et al. 2010). The stakeholder interviews suggested three distinctly different paths that the Hungarian policy community could take to reduce flood losses and provide support to flood victims in the Upper Tisza region: state protection, individual responsibility and ecological development. These paths, which have been discussed extensively in Linnerooth-Bayer et al. (2006), are illustrated in Fig. 12.1 and briefly described below.

12.2.1   State Protection (River Defence)

According to the first path the government continues business-as-usual by taking almost full responsibility for flood risks: building the embankments higher plus generously compensating flood victims. The disincentives for risk reduction are counterbalanced by controlling development in the flood-risk areas. This hierarchical approach – corresponding with today's *river defence paradigm* – was considered unsustainable by some stakeholders insofar as it would likely lead to a worsening of the central government's budget deficit and, despite regulation, encourage undesired development in the flood-prone areas. The interviews and survey revealed that most Hungarians in and outside the Tisza region, however, embraced this state-dominated flood management path, which appears somewhat surprising in light of the current controversies in the Tisza region.

12.2.2 Individual Responsibility

While fiscal necessity forced a partly reluctant government to switch towards more individual responsibility, this switch was positively welcomed by those opposing government intervention more generally. Pro-market opinions emerged from within the private insurance companies (which are foreign-owned to a large degree), as well as the World Bank where a study queried whether protecting crops and homes in the Tisza basin (given most of the farmers are poor) is worth the expense (Halcrow Water 1999). Since economic return from subsistence farming is low, it may make sense according to this voice for the people in the high-risk areas to relocate. Their concern is that government protection and compensation will encourage people to take fewer risk-reducing measures and to move into high-risk flood plains.

Except for insurers and the World Bank, the stakeholders in Hungary hardly attributed responsibility for flood damages to individuals in high-risk areas. There was however many who blamed the new landlords for not maintaining water drains and culverts. Relocation and other individual loss-reducing measures were not popular, nor was there a sense that individuals and communities should be fully insured.

Despite the low opinion of insurance as a workable policy measure, it is remarkable that 60%[2] (40% in the Upper Tisza region) of Hungarian households carried flood insurance offered mainly by foreign-owned insurers (Horváth et al. 2001). The reason for this high uptake was that flood policies were "bundled" with the residential property insurance that is required for a homeowner mortgage. While insurance was not uncommon, insurers offered only limited cover, mainly for breaching or overtopping of the levees. The premium for homeowner flood insurance was independent of the risk; insurers charged all households in Hungary an equal percentage of their property insurance premium (flat rate) to cover flooding. The premiums of those living in high-risk Tisza areas were thus cross subsidized by those living in low-risk areas (for example in large cities like Budapest and Szeged), but still not making insurance affordable to Tisza's poor and vulnerable households.

12.2.3 Ecological Development (Working Landscape)

Critics of both state protection and individual responsibility, many of whom later would support a paradigm shift to a *working landscape*, argued that state protection

---

[2] By 2010 this figure had increased to 72 % (Vereczki 2010).

through the embankment program would further impoverish ecosystems that sustain human and natural systems, and also route flood waters downstream causing damage to the Lower Tisza residents and other riparian countries. Switching to more individual responsibility through insurance was not the answer, however, since this would place an unacceptable burden on an already vulnerable population. In the extreme, it would force people to evacuate the region causing the abandonment of historic villages and residents flooding into unprepared urban areas. Preferred measures would include subsidized programs to help farmers change land-use practices, re-naturalisation of the river by removing embankments in some areas, the construction of retention basins, and provision of infrastructure for soft tourism. Those who were sceptical of this holistic approach pointed out that these measures would not reduce the risks to already existing villages and would exclude commercial insurers from covering floods in Hungary. Nor would it solve the government's budgetary problem.

While views emphasizing state protection, ecological development and to a lesser extent individual responsibility characterized the Hungarian stakeholder discourse, no single stakeholder exclusively advocated any of these policy directions. Almost all stakeholders agreed that levees are important and must remain in many areas and, at the same time, that individuals are responsible for reducing some flood risks. There was unanimity that arguments of economic efficiency are not justified if they increase the burden to an already vulnerable population. Most stakeholders thus supported some government protectionism and, at the same time, saw a limited role for insurance. Still, the policy positions differed greatly. Private insurers put more emphasis on market forces and incentives; national and regional water authorities stressed structural measures and other forms of state protection; and environmental groups gave priority to holistic development and re-naturalization of the watershed. The result of the 3-year policy process was the identification of an intersection of these policy directions, what is called elsewhere a clumsy policy solution (Verweij and Thompson 2006), that commanded wide support (Linnerooth-Bayer et al. 2006).

12.3   Designing a National Insurance Program

Hungary was able to look to other countries as it reflected on its options for a national insurance program, many of which combine public and private responsibility. As one example, the US National Flood Insurance Program (NFIP) offers public insurance that is mandatory for those holding a bank mortgage, and the program is moving from flat-rate to risk-based premiums in order to discourage development in high-risk areas. Alternatively, France's all-hazards insurance system is private but backed by taxpayer funds. In contrast to the US, the French have opted for flat-rate premiums to promote solidarity across regions and hazards. As a further contrast, in the UK insurance is fully privatized and the central government does not provide reinsurance or compensate flood victims. There are many such options that combine solidarity with individual incentives for reducing risks (Linnerooth-Bayer and Mechler 2007).

The Hungarian participatory process began with two information-gathering rounds (stakeholder interviews and public questionnaire) after which the research team proposed three policy paths that appeared consistent with the expressed views. As described below (and also in Linnerooth-Bayer et al. 2006) these paths take account of the apparent widespread stakeholder support for continuing large government involvement in a national insurance program with post-disaster relief to flood victims, as well as the simultaneous endorsement of introducing limited but significant individual responsibility and insurance.

- *Option A*: This business-as-usual option would continue the practice of combining extensive government post-disaster relief with voluntary, flat-rate (cross-subsidized) insurance;
- *Option B*: This option places more responsibility on households living in high-risk areas. The government compensates victims by a lesser amount, and the public role is supplemented by voluntary private insurance based on a flat-rate premium. If a household wishes greater coverage, risk-based insurance would be offered;
- *Option C*: This option reduces the role of private insurers with the creation of a fully public insurance system (government disaster fund) financed by mandatory, flat-rate contributions from all property owners throughout Hungary. The government subsidises insurance premiums for low-income households.

12.4 The Flood Risk Policy Model

To demonstrate the financial consequences of the three insurance options (A, B and C) a flood risk policy model was developed for the Palad-Csecsei pilot region in the Upper Tisza basin in collaboration with VITUKI Consult (this model is described in detail in Brouwers and Riabacke (Chap. 13), this volume. See also, Brouwers 2002; Ekenberg et al. 2003). Depending on the insurance option put into place, the aim of the model was to simulate the incidence of flood losses on three stakeholder groups: flood victims in the pilot basin, the insurance companies and the central government. A second purpose was to examine the usefulness of the model, which was augmented by a graphical user interface, for the deliberations at the stakeholder workshop.

The flood risk policy model simulated the probabilistic distribution of flood losses in the pilot basin over a 10-year horizon and demonstrated the possible effects of selected policy interventions. It consisted of four modules: (1) a one-dimensional, hydrological model of the river based on probabilistic input of water levels at the source, (2) a GIS-based model with values for residential properties, industry and crops in the pilot area, (3) an inundation model with vulnerabilities to physical properties and (4) a policy module that could illustrate the effects of policy changes (see Brouwers and Riabacke (Chap. 13), this volume). The model integrated assessments of the probability of the peril (high water) in the selected geographic region, the probability of levee failure or over-topping of the levee, the vulnerability of the properties concerned and the potential financial loss. The policy module demonstrated the effects of various policy interventions on losses, including, for example, raising the height of the levees or taking them down to create natural reservoirs. The policy model also simulated the effects of insurance arrangements on the profits of insurers, on the government budget and on those living in the pilot basin.

Although the model was designed to be as realistic as possible given available data and knowledge, it was not presented to the stakeholders as full reality. Ravetz (2001) suggests that models be viewed as metaphors, as illustrations of reality without any pretence of representing the full complexity of the physical and behavioural context. Many simplifying assumptions with respect to the data, the scale of the analysis and the functioning of the physical/economic system were necessary. Because of data restrictions on property and crop values, the simulations were carried out for a very small area, the Palád-Csécsei basin, which encompasses only 40 km of the River Tisza and 107 km^2 of flood basin, compared to the county area of 5,937 km^2, or about 2% of the flood area. There are 569,676 inhabitants in the county, of which 4,621 (0.8%) live in the pilot basin. Second, only floods resulting from embankment failures (overtopping or breakage) were taken into account in the simulations. Groundwater related floods were excluded. According to expert judgment, the risk of levee failure was assumed to be salient in only three locations, and levee failures were assumed to result from floods of three magnitudes (100-, 150-, and 1,000-year). There could be at most one levee failure per year. The simulations were based on a 10-year period with 10,000 iterations. To take account of land-use, climate and other changes, the annual flood frequency in this period was assumed to be 10% higher than that calculated from the historical record. The simulated flood damages occurred only to structural property, excluding crop losses and business disruption. The spatial distribution of water depth (but not duration/velocity) was superimposed on the distribution of property values in the pilot area to estimate the direct damages by the use of depth-damage relationships.

The policy module built on the results of the simulated losses in the region and illustrated how these losses would be distributed depending on the choice of a loss-sharing system or national flood insurance program (Ekenberg et al. 2003). In other words, the model addressed "if-then" queries: *If* Option A is adopted, *then* what are the financial consequences to insurers, the government and the local property owners? In an innovative extension of the model, and making use of a stochastic search methodology (see Ermolieva and Ermoliev, Chap. 15 this volume), the flood risk policy model went beyond testing "if-then" questions and generated an "optimal" policy choice based on reasonable assumptions about the preferences of the stakeholders.

The model informed the stakeholder process in a second round of interviews and later at the stakeholder workshop. A slightly different picture emerged from the discussions with the stakeholders when they were informed by the model simulations (see Ekenberg et al. 2003). Most importantly, some key stakeholders realized that the poor could be assisted *before* instead of *after* the flood. This would take the form of taxpayer support for pre-disaster loss reduction measures and insurance. *Government relief could thus be provided through a market mechanism.*

The second round of interviews also showed resistance to incentives for relocating people out of high-risk areas. In fact, many expressed a desire to keep people in the high-risk Tisza area. An anonymous interview with a local mayor revealed this sentiment: "In the Upper Tisza basin, people can survive on very little money and lead reasonable lives, which would not be possible if they were relocated to the cities." Correspondingly, many stakeholders expressed dissatisfaction with instituting risk-based premiums. An exception, not surprisingly, was voiced by an anonymous representative of the Association of Hungarian Insurers (MABISZ), who would like to see more risk-based insurance with the government aiding those who cannot afford the high premiums: "The government should subsidize the poor by the difference between the risk-based and flat-rate premiums."

Making use of the model, stakeholders revised their preferred option to reflect what appeared to be a more moderate support for state protectionism toward more market-oriented and yet pro-poor perspectives. The revisions reflect the almost unanimous view that poor households should be assisted, and the polarized views on the respective roles of private, risk-based insurance and a government fund (for a detailed discussion see Linnerooth-Bayer et al. 2006).

12.5 The Model-Assisted Stakeholder Workshop

The stakeholder workshop was held in September 2002 in the Upper Tisza flood-risk area. Participants included representatives of the key stakeholder groups, including the local mayor, a resident of a non-risk area, the leader of a local environmental group, officials of the regional water management authority and the national authority for disaster management, and a representative of a major international brokerage firm. Unfortunately, the representative from the Hungarian insurance industry was not able to attend (because of a last-minute invitation to attend a meeting on this topic with government representatives). The workshop as a deliberative forum is described in detail in Linnerooth-Bayer et al. (2006) (see also Renn and Thomas 1995; Thompson et al. 1990; Ney 2002). For our discussion, it is important to note the role of the model in aiding the deliberations.

The moderated workshop began with a discussion on flood risk management issues in the region, after which revised options for a flood insurance program (A, B and C) were introduced. The policy model demonstrated how these options distribute flood losses among the three stakeholder groups. The participants were then grouped according to the option chosen and asked to negotiate a common view in their subgroup – reaching a consensus within a single perspective. In what follows we describe the three options that emerged from the stakeholder deliberations.

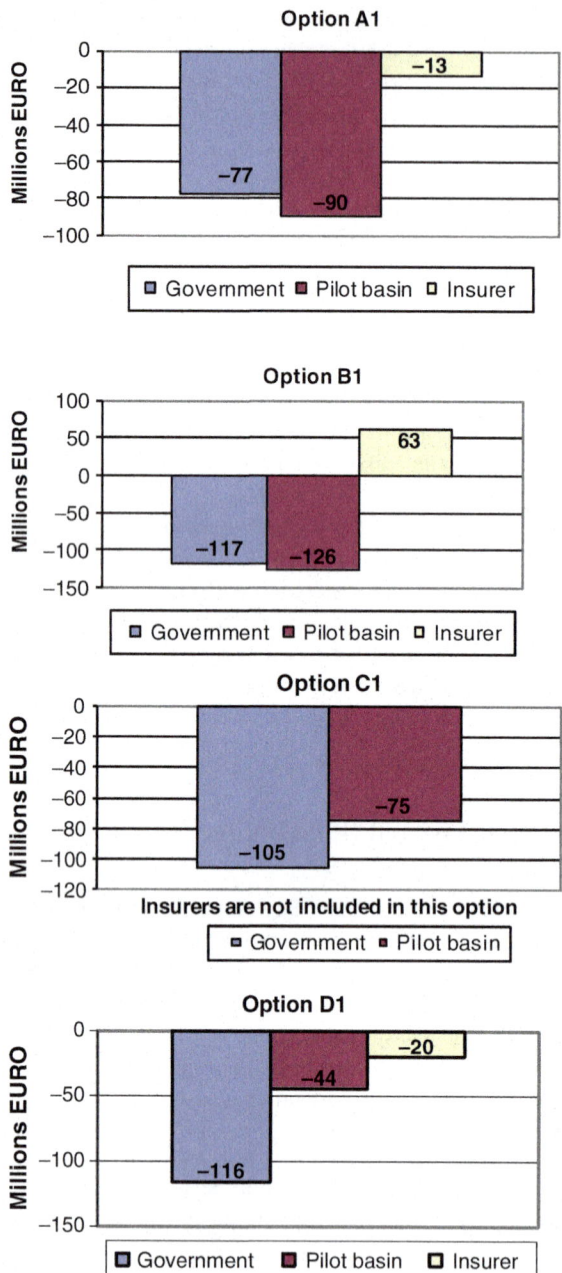

Fig. 12.2 The simulated decadal distribution of losses according to options *A1*, *B1*, *C1* and *D1*

12.5.1   Option A1: A Mixed Public-Private System

The option most closely resembling current practice in Hungary is the mixed public-private system combining insurance with post-disaster government relief (Option A), which was revised by the meeting participants to reduce government compensation and supplement it with voluntary, flat-rate insurance. The spokesperson for the group choosing this option justified this choice as follows:

> We, who were victims of probably the largest flood of the century, feel there is no doubt about the responsibility of the government toward the local population. It wasn't an earthquake or a windstorm you can't be prepared for. In the case of flood disasters the government has a key role and also *has to assume full financial responsibility*. Flood protection lines have been built, and theoretically ... *no water should come out through the dikes*... (emphasis added) (Linnerooth-Bayer et al. 2006)

Instead of incentives for the reduction of risks by residents in the high-risk areas, Group A proposed an interesting innovation: The government and insurance companies would contribute to a fund for financing preventive measures. The final agreement was 50% compensation to flood victims from the government and the rest would be dependent on the voluntary purchase of insurance, as detailed in the box below:

Option A1: A Mixed Public-Private System
- Government compensation to private flood victims for 50% of their damages;
- A private insurance system with

  - voluntary policies (uptake rate assumed to be 50%),
  - bundled or separate policies for all types of natural disaster risks, covering up to 75% of the damage,
  - no deductibles, and
  - flat-rate premiums (calculated as 0.015% of property value);

- Government subsidies for poor households up to 100% of premium;
- Reinsurance from the private market; and,
- Insurers contribute to a prevention fund (not included in the model).

12.5.1.1   Model Simulations for Option A1

Figure 12.2 shows the results of the simulated financial consequences of all three options, plus the negotiated compromise (Option D) for the government, the insurance policy holders and the insurers. The simulations for Option A1 are based on the parameters shown in the above box, and also on a number of clarifying and simplifying assumptions. It was assumed that 50% of households in the region would purchase insurance cover. Consistent with government estimates, it was also assumed that 60% of the households in the pilot area are poor, which means large subsidies from the government to poor households for purchasing insurance.

Finally, the insurers operate with no transaction costs. Because of these limiting assumptions, the simulation results should be regarded as only rough approximations of the financial burdens, at most indicating the relative burdens among the three stakeholder groups.

The simulated distribution of losses from Option A1 totaled €180,000 for the decade. The government continues to absorb a large percentage of the losses, and the insurers can also expect a net loss. It is clear that the insurers would not agree to this system, which closely resembles current practices, and may help explain why insurers are pulling out of high-risk regions.

12.5.2 Option B1: Private Responsibility and Insurance

Turning to Option B1, significantly more reliance is placed on private insurance than is currently the case; however, to protect poor households the government subsidizes insurance premiums for low-income households. This option has the advantages of discouraging (non-poor) persons from locating in high flood-risk areas and placing the burden on the pool of exposed persons rather than on taxpayers. Given weak Hungarian support for individualistic policy paths, it is not surprising that only one participant at the workshop selected Option B1 as the preferred policy, and justified this choice as follows:

> We have to consider that if the government covers all the costs, this actually means *that the costs are covered from the pockets of the citizens,* including us... Therefore, I agree with the proposal that the government should not directly provide compensation. ... (emphasis added) (Linnerooth-Bayer et al. 2006).

The government should not provide compensation, but taxpayers are called upon in two ways. Poor households can purchase subsidized flood insurance, and a government catastrophe fund (a type of reinsurance) would cover claims in the case of very large or multiple catastrophes that go beyond the means of insurers. The details of Option B1 are listed in the box below:

Option B1: Private Responsibility and Insurance
- No post-disaster government compensation to private flood victims;
- Private insurance system characterized by
    - non-mandatory policies,
    - cover for flood, standing water, and all hazards (not modelled),
    - insurance purchased separately or bundled with property insurance policies,
    - up to 100% cover for damage without deductibles, and
    - risk-based premiums;
- Government subsidies for poor households up to 100% of premium;
- Government re-insurance fund financed by tax revenues.

12.5.2.1 Model Simulations for Option B1

What is notable about Option B1 (see Fig. 12.2) is the relatively advantageous outcome for private insurers, and no doubt insurers would have preferred this option if they had participated in the workshop. The government has large expenditures because of its role in subsidizing low-income households and providing a guarantee for the private insurers. The model showed that with 90% probability, the government losses over the decade are over €40,000 (this is relatively small, but it should be kept in mind that this exercise covers only the small pilot basin). Since losses to residents in the pilot region are averaged over those with and without insurance cover, these losses appear equivalent to Option A1; however, this masks the equity issues since uninsured households (40%) will be devastated in this scenario. A question, which was not dealt with by the stakeholders, is whether the government would not then be under extreme political pressure to provide aid to uninsured households? For the representative insured household, the loss in flood years is only the insurance premium of €32. This annual premium is risk based, and is equivalent to about 0.07% of the selected, high-property value household.

12.5.3 Option C1: A Public Insurance Fund

According to Option C1, the government acts as the primary insurer, thus eliminating private insurance as an option. All property owners throughout Hungary would be required to purchase catastrophe insurance from the government. The premium will not be based on the locational risk of the household but will take the form of a flat-rate tax on property. This will shift liability for disaster reconstruction from taxpayers to property owners. Private insurers administer the system on a commission basis by collecting premiums and by assessing and paying post-disaster claims. The premiums contribute to a public catastrophe fund; however, if premium income is not sufficient to cover losses, taxpayers will be called upon to supplement the fund in addition to subsidizing poor households (up to 100%) in their purchase of insurance. Four participants at the stakeholder workshop selected this option. The spokesperson for this small group justified the choice as follows:

> In Hungary, the floods have already shown that *there are risks for which insurance companies are unwilling to offer coverage on a market basis*, and that's why we need a catastrophe fund...operated according to non-market principles ... but not necessarily funded by the central budget. (emphasis added) (Linnerooth-Bayer et al. 2006).

In discussing this option, it was emphasized that risk-based pricing would be complicated (especially if the system is expanded to include other hazards) and also would exclude the poor and highly vulnerable. The eventual agreement on the details of the system by Group C are shown in the box below:

Option C1: Mandatory Public Insurance with Government Subsidies
- A public insurance system administered by private insurance companies with

  - mandatory policies (100% uptake is assumed),
  - cover extending to all natural disasters (not modeled),
  - cover extending up to 90% of the damage,
  - no deductibles, and
  - fund covered by a flat-rate tax (0.02%) on all property owners;

- Government subsidies to poor households up to 100% of premium;
- Government underwrites all risks including the risk of diverting the catastrophe fund.

12.5.3.1 Model Simulations for Option C1

As shown in Fig. 12.2, the flood risk policy model simulated the financial consequences of Option C1 for those directly affected by floods in the pilot region – in this case only the government and resident households. Since the model was unable to take account of additional risks that would be covered by an all-hazards insurance system, the simulations reflect only the loss-sharing consequences based on flood risks. It was assumed that property owners pay a special flood tax equivalent to 0.02% of their property value to the government catastrophe fund. This is roughly equivalent to what households currently pay for flood insurance, which results in significant cross subsidies across Hungarian (insured) property owners.

Looking at Fig. 12.2, this scenario is clearly superior to that of Option B1 in the sense that both the government and residents can expect lower losses. The reason, of course, is the absence of profits for the insurers. In a flood-free decade the government fund will have a positive balance by collecting the flood tax from all property owners in the pilot basin. In decades when floods occur, however, the costs to government are substantial since the government awards close to full compensation to the flood victims.

12.6 A Consensus Policy Path

After arguing for their policy perspectives, the workshop participants turned to a lively and heated discussion on a possible compromise, which led to an imaginative new system (Option D, Fig. 12.2): Only *households with private insurance would qualify for government assistance after a disaster, but the government would*

Table 12.1 Ranking of option D for the stakeholder groups under different decadal conditions

Stakeholder groups	Simulated decade	No-flood decade	Severe-flood decade	Non-weighted average
Government	Middle (2)	Worst (1)	Middle (2)	1.7
Pilot Basin Residents	Best (3)	Middle (2)	Best (3)	2.7
Insurers	Worst (1)	Best (3)	Worst (1)	1.7

heavily subsidize poor households in their purchase of voluntary, private flood insurance. It was also agreed that the government would not provide reinsurance for private insurers. The details are shown in the box below:

Looking at Fig. 12.2, we see that the government continues investing significantly in post-disaster assistance in the Tisza area, but its simulated losses are somewhat less than for Option B1. The additional expense of subsidizing poor households means that the government's fiscal problem is not solved, at least not in this decade (and the model demonstrated this).

Consensus Option (D)
- Government compensation only to insured households (50% of losses);
- A private insurance system with

    - bundled or separate policies for all types of natural disaster risks (not modelled),
    - covering approximately 50% of the damage,
    - voluntary, flat-rate premiums (0.01% of property value),
    - no government reinsurance;

- Government subsidies for poor households up to 100% of premium.

12.6.1 Comparing Options

Table 12.1 shows the best, worst and mediocre options for the government, residents of the pilot basin and the insurers for the simulated decade, as well as for decades with "no floods" and with "severe floods". For the simulated decade, the pilot basin residents stand to gain the most from the consensus option, and also for a decade with severe floods. For the government, the picture is very different, and Option D is never the "best" option. The government stands most to lose in a no-flood decade (because it subsidizes insurance premiums for poor residents). The consensus option is the worst for insurers for the simulated decade, and insurers only stand to gain in the unlikely case of no floods.

Overall, the pilot basin residents, but not the government, stand to gain from this consensus solution. Why, then, did the government representatives support this system? As one of the participants remarked, "this policy will cost the government dearly in the short term, but it will create a culture of responsibility and insurance in the region over the medium term" (Linnerooth-Bayer et al. 2006). Indeed, the government representatives saw Option D as shifting responsibility to individuals over a long horizon, or building a culture of individual accountability. They acknowledged the difficulty of imposing large burdens on a vulnerable and distrusting population overnight. It is also interesting that the environmentalists, who had originally been opposed to insurance, supported this gradual switch. The heavy subsidization of insurance premiums for poor households placated those who have a moral commitment to pro-poor policies and social justice. Interestingly, during this 3-year process, trust in insurance companies appeared to increase, perhaps with recognition that most insurers in Hungary are foreign based and financially more secure than their Hungarian counterparts.

While it is remarkable that a consensus (absent insurers) was reached, many caveats are in order. The nine persons at the workshop may not have adequately represented the full range of stakeholder perspectives and interests. The solution on which the stakeholders decided would not have been endorsed by insurance companies, as it would have required them to offer greatly expanded cover at flat-based rates. It is regrettable that the high-level representative from the insurance industry cancelled his invitation to the stakeholder workshop to attend a meeting at the Finance Ministry.

12.6.2 The Hungarian Insurance Legislation

Following the stakeholder workshop, and partly influenced by the results, the Hungarian government decided upon a flood insurance program for properties threatened by damages from riverine flood or standing water. According to legislation in 2003 (the so-called Wesselényi Miklós Fund (Act LVIII.)), the government has fully underwritten flood insurance in high-risk areas where most private companies do not offer policies.[3] The program is administered by the regional offices of the Hungarian Treasury and some local governments, and the government provides a back-up if the premium pool is insufficient to cover claims. In direct contradiction to the results of the stakeholder compromise, insurance premiums will be risk-based, and the premiums of poor households will be subsidized only up to 30%.

In other respects the legislated program is consistent with the stakeholder solution: Insurance is voluntary and available only for homes built with a permit. It covers all flood and standing water damages both in protected and unprotected

---

[3] No private insurance company has offered insurance against the risk of standing water that is not related to riverine flood.

flood basins. The indemnity can reach 100% of the property value (in contrast to 50% as agreed by the stakeholders), but with a maximum of Euro 57,000.

To date uptake of this insurance is low. In 2010 the number of properties insured within the framework of the Wesselényi Miklós Fund was 876, while the number of entitled property owners was over 100,000 (Vereczki 2010). Its Achilles heel seems to be that poor households will only receive a 30% subsidy for their private, risk-based insurance premiums. The great majority of Hungarians still feel that the government should be responsible for flood prevention and compensation, and in case of severe floods the government still compensates flood victims. For example, during the 2010 summer floods and storms, when the estimated losses of the households totalled around 143 million Euro,[4] government subsidies reached around 11 million Euro (Vereczki 2010).

12.7 Conclusions

This book is dedicated to the notion that the development of efficient and equitable policies for managing disaster risks and adapting to global environmental change is critically dependent on robust decisions supported by integrated modeling. This chapter has demonstrated how an integrated model can support a stakeholder process in the design of a national flood insurance program in Hungary. The Tisza stakeholder process, which combined information technology with public participation through stakeholder interviews, surveys and a workshop, was innovative in many ways: It was uniquely based on the methodology of catastrophe modeling by combining Monte Carlo simulations to generate the incidence of flood losses taking account of different options for an insurance program. The model helped to clarify how simulated flood losses would be shared among the victims, the government and the insurers depending on the design of the insurance pool. Model-based techniques for public involvement were grounded in a recognition and respect of diverse values and worldviews on the part of the stakeholders and the need to reach a consensus mindful of this diversity. Finally, the stakeholder process reached a pilot consensus on an innovative disaster insurance scheme for Hungary, which was influential in the legislation that was adopted following publication of the study.

The resulting solution based on only nine workshop participants clearly cannot claim to be representative of the full policy terrain in Hungary; in fact, the insurance company voice was underrepresented at the workshop. The value of this pilot study lies rather in pointing to a new form of policy analysis that makes use of information technology and concepts of consensus policy solutions in participatory, stakeholder settings, and is respectful and mindful of the different views of the problem and solutions. It is hoped that the methodology and model-based participatory

---

[4] This figure includes damages caused by both floods and storms.

processes presented in this chapter will be useful for all researchers intent on providing useful information to controversial policy issues such as those characterizing the policy discourse today in Hungary.

References

Bran J, Borsos B (2009) Conservation and restoration of the globally significant biodiversity of the Tisza river floodplain through integrated flood plain management. Report commissioned by the Hungarian Ministry of Environment and Water, Government of Hungary and United Nations Development Program, UNDP Atlas ID 0046904

Brouwers L (2002) Spatial and dynamic modeling of flood management policies in the Upper Tisza, Interim report 03–002. International Institute for Applied Systems Analysis, Laxenburg, Austria

Ekenberg L, Brouwers L, Danielson M, Hansson K, Johansson J, Riabacke A, Vári A (2003) Flood risk management policy in the Upper Tisza basin: a system analytical approach: simulation and analysis of three flood management strategies, Interim report 03–003. International Institute for Applied Systems Analysis, Laxenburg, Austria

Hajós B (2002) Flood defense in Hungary in the 21th century. Ezredforduló 2:24–27 (in Hungarian)

Halcrow Water (1999) Flood control development in Hungary: feasibility study final report. Halcrow Group Ltd., London

Horváth G, Kisgyörgy S, Sendzimir J, Vári A (2001) The 1998 Upper Tisza flood, Hungary: case study report, Draft paper. International Institute for Applied Systems Analysis, Austria

Linnerooth-Bayer J, Mechler R (2007) Disaster safety nets for developing countries: extending public–private partnerships. In: Amendola A, Linnerooth-Bayer J et al (eds) Special issue on financial vulnerability. Environ Hazards 7 (1):54–61

Linnerooth-Bayer J, Vári A, Thompson M (2006) Floods and fairness in Hungary. In: Verweij M, Thompson M (eds) Clumsy solutions for a complex world: governance, politics and plural perceptions. Palgrave Macmillan, New York

Ney S (2002) Focus groups, citizen participation and governance in Europe, Synthesis report for EU project on pension reform and citizen participation. ftp://ftp.iccr.co.at/penref/d4-synthesisreport.pdf

Ravetz J (2001) Models as metaphors: a new look at science. The Research Methods Consultancy Ltd., London

Renn O, Thomas W (1995) A brief primer on participation. Philosophy and practice. In: Renn O, Webler T, Wiedemann U (eds) Fairness and competence in citizen participation: evaluating models for environmental discourse. Kluwer, Dordrecht

Sendzimir J, Pahl-Wostl C, Kneiper C, Flachner Z (2010) Stalled transition in the upper Tisza river basin: the dynamics of linked action situations. Environ Sci Policy 13(7):604–619

Thompson M, Richard E, Wildavsky A (1990) Cultural theory. Westview Press, Boulder

Vári A, Linnerooth-Bayer J, Ferencz Z (2003) Stakeholder views on flood risk management in Hungary's Upper Tisza basin. In: Linnerooth-Bayer J, Amendola A (eds) Special edition on flood risks in Europe. Risk Anal 23:537–627

Vereczki A (2010) Insurance and extreme weather. "CLIMA-21" brochures. Clim Change Impacts Responses 63:73–82 (in Hungarian)

Verweij M, Thompson M (2006) Clumsy solutions for a complex world: governance, politics and plural perceptions. Palgrave Macmillan, New York

Chapter 13
Consensus by Simulation: a Flood Model for Participatory Policy Making

Lisa Brouwers and Mona Riabacke

Abstract An overall goal of the Upper Tisza flood risk management project was to design a flood management policy that shared liability for disaster losses between the central government and individual households in a way that was considered acceptable by all the stakeholders. A participatory approach was adopted, where a flood simulation model was used interactively to support the process. In this chapter, we describe the design, implementation and use of the dynamic and spatially explicit flood simulation model, which incorporated novel elements like micro-level representation and Monte Carlo techniques. The model was provided with an interactive graphical interface designed to facilitate its use as a decision support tool in a participatory setting with multiple users. During this process, the model supported comparisons between pre-defined policy options, as well as the design of a new policy option on which consensus was finally reached.

Keywords Catastrophe modeling • Decision Support tool • Flood risk management • Flood simulation model • Stakeholder processes • Tisza

13.1 Introduction

Risk and environmental policy making is characterized by multiple subjects with different values, knowledge, perceptions and interests: this can make the decision process long and complex. In democratic societies, participatory procedures are

L. Brouwers (✉)
School of ICT, SCS, KTH Royal Institute of Technology,
Forum 120, SE-164 40 Stockholm, Kista, Sweden
e-mail: lisabrou@kth.se

M. Riabacke
Department of Computer and Systems Sciences,
Stockholm University, Forum 100, SE-164 40 Kista, Stockholm, Sweden
e-mail: mona.riabacke@gmail.com

A. Amendola et al. (eds.), *Integrated Catastrophe Risk Modeling: Supporting Policy Processes*, Advances in Natural and Technological Hazards Research 32, DOI 10.1007/978-94-007-2226-2_13, © Springer Science+Business Media Dordrecht 2013

increasingly promoted as a way to reach consensus through stakeholder deliberation (Stern and Fineberg 1996). Recent advances in information technology can provide citizens with relevant information and facilitates their participation in the political processes in what is now called "eDemocracy" (Macintosh and Whyte 2006). The use of decision support systems and decision analytical techniques during participatory policy-making processes has proved beneficial to both the decision process and the acceptance of resulting decisions (Gregory et al. 2005; Rios Insua et al. 2007). Indeed, these techniques offer a common platform, where different viewpoints can be modelled and compared (Jiggins and de Zeeuw 1992; Cain et al. 2003; Jakeman et al. 2006). In the case discussed in this chapter, the model facilitated discussion of pre-proposed policy options for flood risk management by illustrating and clarifying their impacts on the different stakeholders. In so doing, the model evaluated different flood scenarios or narratives describing alternative futures.

Models can be developed at different levels of resolution. When dealing with decisions like flood management affecting communities, a micro model, where each individual or household is represented explicitly, has the advantage that it can investigate the distributional effects of different policy options. Average outcomes might hide inequalities. A policy that seems acceptable on the average can be disastrous to certain individuals. Fairness and equity among individuals, societal groups and/or geographical regions are factors that should be considered in policy design (Linnerooth-Bayer and Amendola 2000).

Typically the outcomes of policy options depend on uncertain variables; this is the case when dealing with risk of flooding or other natural disasters. Modern computer techniques use Monte Carlo simulations to explore scenarios taking account of uncertainty. In each simulation the uncertain variables are assigned new random values within their distribution, and the process is repeated a large number of times to achieve statistically reliable outcomes. During the last two decades, spatially explicit catastrophe models have been used primarily by insurance and re-insurance companies for quantifying the risk of damage exposure (Walker 1997; Ermolieva 1997; Amendola et al. 2000a, b; Ermolieva and Ermoliev 2005). These models typically cover large geographic areas and use large amounts of property data and land-use data for calculating the effects of simulated events. Due to the data intensity, the models are often aggregated by region or zip code area. Therefore, fine distributional effects of imposing different catastrophe policies are lost.

In the Upper Tisza flood risk management project, the overarching goal was to design a socially and economically acceptable national insurance system that would shift part of the economic post-disaster liabilities from the central government to individuals (Ermolieva et al. 2003; Vári et al. 2003). This chapter describes the design and implementation of a flood simulation model that was used to support the stakeholders in the identification of acceptable flood risk insurance policy options in a participatory manner. The use of the model in the participatory process is described in more detail in Chap. 12 by Linnerooth-Bayer et al.

The flood simulation model combined the features of micro models and catastrophe models. The simulations are spatially explicit and make use of disaggregated data on the level of households. Conceptually, it can be divided into four modules: the disaster module, the policy module, the consequence module, and the interface module. In Sect. 13.2, Simulation Model, we present the theoretical background of the simulation model, followed by a conceptual description of how the different modules interact. In Sect. 13.2.1, Disaster Module, we describe how the catastrophic flood events were generated and in Sect. 13.2.2, Policy Module, we present how different policy options were constituted and how they were tested by the model. Section 13.2.3, Consequence Module, provides a description of the economic update rules for all agents. Section 13.3 Interface Module, describes the graphical user interface and the principles that governed its design. In the concluding Sect. 13.4 we describe how the simulation model was used during the concluding stakeholder workshop (a full description can be found in Linnerooth-Bayer et al. at Chap. 12 of this book).

13.2 Simulation Model

The flood management issues that were to be represented in the model were identified through interviews and surveys with the involved stakeholders, in this case, the government, the insurance companies, the individual property owners, and the regional authorities. The importance of eliciting stakeholder values as a basis for model design has been pointed out (Friedman 2004). In the Tisza flood management case, following the initial interviews (Vari et al. 2009; Linnerooth-Bayer et al. 2006) significant economic indicators were identified that represented the outcomes most important to the four stakeholder groups. To ensure that comparisons between different policies were possible, it was important that all stakeholders had a common frame of reference, in this case monetary outcome. In a participatory setting, the outcome is ideally presented from different stakeholder perspectives.

A simulation approach was adopted to account for the complex and uncertain system context. There were uncertainties on the macro-level (flooding intensity and related levee reliability) as well as on the micro-level (distribution of insurance contracts and of economic wealth). The uncertainties in combination with the dynamic property of the policy model made the space of possible outcomes very large. In policymaking, micro-level models are increasingly used (Mitton et al. 2000). Since micro data on every household in the pilot region was available, it was decided to keep the data disaggregated in the model to allow for detailed analyses. Only financial policies were incorporated in the model since estimations of costs and consequences of imposing other types of policies were beyond the scope of this pilot project.

The simulation model consisted of four modules: the disaster module, the policy module, the consequence module and the interface module. The flood events were generated by the disaster module. The flood management policy was specified in the policy module. The economic outcomes for the stakeholders were updated annually in the consequence module. The interface module provided the users

220 · L. Brouwers and M. Riabacke

Fig. 13.1 Hungary with the
Szabolcs-Szatmár-Bereg
County

with a graphical interface to enable interactive involvement and to display the
outcomes graphically. The basic time-step in the simulation model was a year, and
the time frame in these experiments was the forthcoming 10 years. Since the model
included randomness, an experiment consisted of a large number of simulations to
guarantee stable results.

13.2.1   Disaster Module

The river basin, located in the Szabolcs-Szatmár-Bereg County is divided into 11
municipalities, see Fig. 13.1.

Neighbouring countries are Romania, Slovakia and the Ukraine. In the model,
the basin was geographically represented in the form of a grid, consisting of
$1{,}551 \times 1{,}551$ cells of 10 sq. m each. There were 2,580 private properties in
the basin. Since this part of the Tisza River is protected by flood levees, only
levee failures are considered as disaster events. A levee failure occurs when a levee
breaks or is overtopped. Hydrologists at Vituki Consult Rt. calculated the
probabilities for nine levee failure scenarios that could occur in the studied part
of the river. The nine scenarios were based on the assumption that a levee failure
could occur at one of three river locations. It was also assumed that a flood had one
of three possible magnitudes (100-, 150-, or 1,000-year flood). The combination of
location and magnitude resulted in nine failure scenarios, referred to as different
flood states of the system. The tenth and complementing state was the zero-event
when the levees did not fail and thus no disaster occurs.

The probabilities for the levee failure at three locations are presented in
Table 13.1. Monte Carlo techniques were used to simulate the annual flood state
of the system. The variable flood was assigned a random number in the range [0, 1]
from a uniform distribution. The value was checked against nine threshold values.
If the value is equal to or smaller than 0.0012, scenario 1 occurs (failure at location
1 from a 100-year flood), if it is smaller than or equal to 0.0032 scenario 2 occurs
(failure at location 1 from a 150-year flood), and so on. If the value is greater than

Table 13.1 Levee failure probabilities

Location	1	2	3
Levee failure from 100 year flood	0.12	0.20	0.28
Levee failure from 150 year flood	0.18	0.22	0.40
Levee failure from 1,000 year flood	0.19	0.33	0.45

Source: Vituki (1999)

the last threshold (0.0123) scenario 10 (no failure) occurs. The failure probabilities were obtained through the compound probability of a flood and the failure of the levee. For instance the probability for the event, levee failure at location 3 from a 1,000-year flood, is 0.00045 (0.001 × 0.45).

13.3   Policy Module

Since it would be impossible to consider all possible flood management policies, even those associated with a national insurance program, a subset of the most relevant policy parameters was identified. By varying the values of these parameters in the simulation model it was possible to investigate selected policies. The same approach was used by the Intergovernmental Panel on Climate Change (IPCC) when they developed a set of long-term emission scenarios describing how greenhouse gases emissions could evolve between the years 2000 and 2100 (Intergovernmental Panel on Climate Change (IPCC) Working Group III 2000). The policy options in this flood management study were based on the views elicited by the initial stakeholder interviews and surveys as described in Chap. 12 (Linnerooth-Bayer et al.). They reflect the widespread stakeholder support for continuing large government involvement in a national insurance program with post-disaster relief to flood victims, as well as the simultaneous endorsement of introducing individual responsibility and insurance. The three pre-defined policy options that were investigated in the simulation model at the stakeholder workshop (as described in Linnerooth-Bayer et al. Chap. 12) are briefly summarized below.

Option A1: A mixed public-private system

- Government compensation to private flood victims for 50% of their damages;
- A private insurance system with

   - voluntary policies (uptake rate assumed to be 50%),
   - bundled or separate policies for all types of natural disaster risks,
   - covering up to 75% of the damage,
   - no deductibles, and
   - flat-rate premiums (not differentiated by risk class) calculated as 0.015% of property value,

- Government subsidies for poor households up to 100% of premium;
- Reinsurance from the private market.

Option B1: Private responsibility and insurance

- No post-disaster government compensation to private flood victims;
- Private insurance system characterized by

  - non-mandatory policies,
  - cover for flood, standing water, and all hazards (not modelled),
  - insurance purchased separately or bundled with property insurance policies,
  - up to 100% cover for damage without deductibles, and
  - risk-based premiums;

- Government subsidies for poor households up to 100% of premium;
- Government re-insurance fund financed by tax revenues

Option C1: Mandatory public insurance with government subsidies

- A public insurance system administered by private insurance companies with

  - mandatory policies (100% uptake is assumed),
  - cover extending to all natural disasters (not modeled),
  - cover extending up to 90% of the damage,
  - no deductibles, and
  - fund covered by a flat-rate tax (0.02%) on all property owners;

- Government subsidies to poor households up to 100% of premium;
- Government underwrites all risks including the risk of diverting the catastrophe fund

As described in Chap. 12 the purpose of the stakeholder workshop was to explore the potential for a compromise option for a national flood insurance program. The workshop deliberations were supported by the simulation model that facilitated the interactive design of new options by allowing the workshop participants to change parameters, such as insurance premiums and government compensation, and simulating the economic impacts of each new option on the four stakeholder groups.

13.4 Consequence Module

The economic outcome of alternative policy options were presented from four stakeholder perspectives: government, insurer, property owner and regional administration (the aggregated economy of all property owners in the pilot basin). The economic outcomes were saved after each model run, that is, over a simulated 10year period. After a sufficient number of simulations e.g. of 1,000 runs, the outcomes of a policy experiment from all runs were analysed.

The outcome from a single run was not merely the result of the flood states and flood compensation policy over the 10-year period; for each simulation run it was randomly decided if individual property owners held an insurance contract or not.

The overall proportion of insured properties was either decided beforehand in the three policy options, see Sect. 12.2.2 above, or it was decided interactively through the user interface. There were 2,580 properties, and thereby over six million possible combinations of the binary insurance choice (insured/not insured).[1] The insurance distribution stayed fixed during a single run, that is, household insurance choices did not change during the 10-year period.

According to statistics, as much as 60% of the households in the region were classified as poor (KSH 2000). In the simulation model, this affected the extent to which the government subsidised insurance premiums for low-income property owners. Since we did not have access to micro data on household income, we reduced the income variability into two states: poor or non-poor. The income distribution stayed fixed during a simulation run. Being poor did not affect the likeliness to buy insurance since according to all selected policy options the government subsidized premiums.

13.4.1   Damages

For each of the ten levee failure scenarios (nine with failures and one without) the model estimated a corresponding probabilistic distribution of economic damage to each property. The total damages from a levee failure scenario were aggregated in each of the 11 municipalities.

The expected annual damage for an individual property owner determined the risk-based insurance premium. It was based on the average risk of the municipality, that is, the total expected damage per municipality divided by the number of properties.

13.4.2   Governmental Wealth

The extent to which the government subsidises the insurance premium for a poor household was determined beforehand in the three pre-defined policy options or by the users before starting a new experiment. After a levee failure, the government compensated the owners of the flooded properties based on the option-specific predetermined percentage of their damages. The economic balance of the government is described in Eq. 13.1 in which *GovSubs* represents subsidised insurance premiums and *GovComp* represents compensation for flooded properties and *t* denotes time (the current year of simulation).

$$GovBalance(t) = GovBalance(t-1) - GovSubs(t) - GovComp(t) \qquad (13.1)$$

---

[1] $2{,}580^2$.

13.4.3   Insurance Company

The insurance company receives income from insurance premiums. The premium depends on the property value, the coverage (the fraction of property that is insured) and the differentiation of the premium (flat-rate or a risk-based). As defined earlier, a *flat-rate premium* depends on the property value together with the size of the coverage. For example, if the premium is 0.1% of the property value, and the coverage is 80%, then the annual premium for a property worth 200,000 USD is 160 USD. In contrast, a *risk-based premium* takes into account the expected *local* damages (the average expected damages in the municipality) multiplied by the insurer's add-on (safety loading) and the size of the coverage for that property.

The insurer compensates policy holders who suffer flood damage depending on the extent of the damage and the level of the coverage. Assuming damage of 10,000 USD, a policy 70% coverage would entitle the policy holder to 7,000 USD compensation. Equation 13.2 describes the accounting book of the insurer. *PremFR* denotes premium income from flat-rate premiums and *PremRB* from risk-based premiums; *Comp* denotes the paid compensation.

$$
\begin{aligned}
InsBalance(t) = {} & InsBalance(t-1) + PremFR(t) + PremRB(t) \\
& - Comp(t)
\end{aligned}
\tag{13.2}
$$

13.4.4   Property Owner (Individual and Aggregated)

Property owners classified as poor are entitled to subsidized premiums to the degree specified in the policy option. Equation 13.3 describes how the economic balance of the property owners is calculated. The household economics or balance is reduced by flood damages, *Dam*, and insurance premiums. Governmental subsidies and damage compensation increase the balance.

$$
\begin{aligned}
PropBalance(t) = {} & PropBalance(t-1) - Dam(t) - PremFR(t) \\
& - PremRB(t) + GovSubs(t) + Comp(t)
\end{aligned}
\tag{13.3}
$$

The aggregated balance for all property owners, the *Pilot Balance*, is described in Eq. 13.4.

$$
PilotBalance(t) = PilotBalance(t-1) + \sum PropBalance(t)
\tag{13.4}
$$

For each of the four stakeholder perspectives, the policy-relevant parts of the wealth described in Eqs. 13.1, 13.2, 13.3, and 13.4 were updated annually.

13.5 Interface Module

The complexity of the micro simulation model made it unsuitable for direct use by the workshop participants, who would supply input to the model (by designing policy options) and interpret the output (by using the visualization as a basis for discussions). In order to be suitable as a decision support tool in the participatory setting with multiple users, the model was provided with an interactive interface. This would enable the interactive discussion and revision of policy variables, e.g., how varying degrees of compensation would impact the four stakeholder groups.

Usability was an essential aspect, and a requirement for the user interface was that all users with the relevant background knowledge should be able to interact with the model regardless of their computer skills. It was important to engage all participants in a limited time; therefore an interface that was easy to learn in a short time was prioritized. Ideally, the economic consequences for the different policy options would be presented taking into account as many of the potential stakeholder objectives as possible, such as, consequences for the government, the insurer, the individual, and the region as a whole. In addition, it was important that the results were displayed such that misconceptions regarding the importance of individual parameters could be refuted.

A main consideration for the design of the graphical user interface was which variables from the model should be made available to users. The pros and cons of exposing each model variable had to be weighed: hiding a variable at the cost of compromising transparency or making it accessible at the risk of complicating the use of the tool. The policy variables selected for display included, for example insurance rate, premium and level of government compensation. To simplify the input procedure and make it more suitable for collaborative work, the interaction with the model was accomplished through mouse input, choosing values from pop-up menus, dragging sliders or choosing radio-buttons. Results from similar projects, for example, the ULYSSES project (Dahinden et al. 2000) show that such input procedures are preferable to participants. Also, these input procedures increase process awareness for the whole group (and not just for the person in control of the keyboard) as everyone can follow the course of events and view the selected options and their impacts. The different steps of the simulation procedure suggested a natural division of interaction with the model into three stages: (1) *choose a mode*, (2) *set variables*, and (3) *view results*. These stages are described below.

13.5.1 The Main Window: Choose a Mode

The tool can be used in two separate modes: the *analyze mode* (analysis of three pre-defined policy options, where some of the parameters can be changed by the user) or the *experiment mode* (design of new policy options). Figure 13.2 shows the main window of the tool, where the user can choose to open the "settings" window of either one of the two modes.

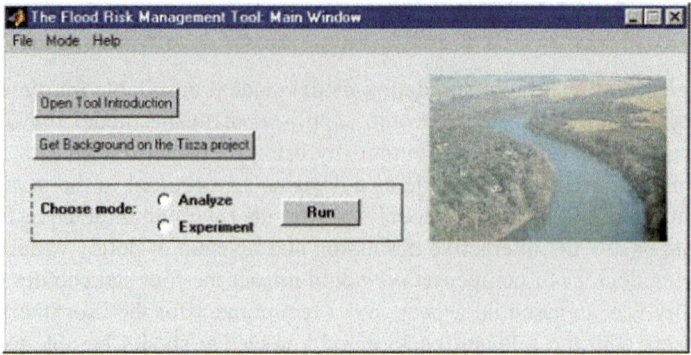

Fig. 13.2 The main window

13.5.2 Settings Windows: Set Variables

When simulating policy options, the user has the option of choosing the time frame (number of simulated years), the number of simulation rounds in an experiment and whether to increase/decrease the flood frequency rate.

Each mode has a specific purpose. In the analyze mode, the user can explore variations of the three proposed flood management policies (options A1, B1 and C1), here referred to as scenario 1, 2 and 3. In the settings window of the analyze mode (see Fig. 13.3), the user can choose which scenario(s) to include in the simulation. The user can choose the insurance rates (the percentage of households with insurance) of *insurance 1* (scenario 1 and 2) and *insurance 2* (scenario 2).

The experiment mode was designed to support explorative processes where users could modify existing policies and design new ones (up to three at a time). In the settings window of the experiment mode, the users design their own policy options by setting a group of parameters, namely *yearly income, insurance rate 1, insurance premium 1, compensation from insurance 1, insurance rate 2, compensation from government, government acts as re-insurer* and *flood tax*.

13.6 Results Window: View Results

The results are displayed in a common view for all simulated policy options (A1, B1 and C1, or new ones). If three scenarios are simulated, the user has the option to switch between displaying the results of the multiple scenarios or between the four stakeholder perspectives.

Results corresponding to each perspective are displayed in two graphs, see Fig. 13.4. The top graph shows the different economic outcomes resulting from the simulation, whereas the lower graph shows the corresponding percentage occurrence of each outcome, that is, the number of times that each outcome

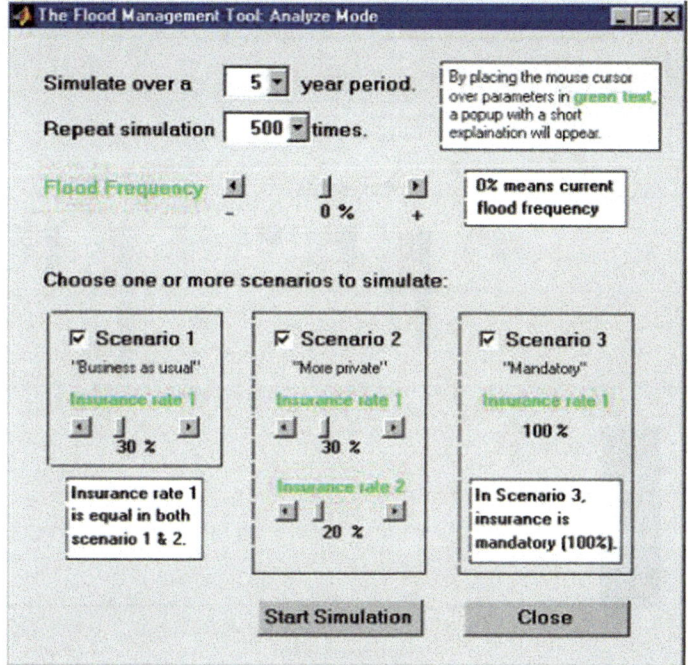

Fig. 13.3 The settings window for the Analyze mode

occurred out of all simulations. This view enables the user to perform a relatively easy comparison of different policy options, for example, in the majority of the outcomes (95.2%) the government had 0 HUF[2] expenses in scenarios 1, 2 and 3, but in the most extreme case (0.2% of the outcomes) the government had a loss of over 400 million HUF for scenario 2, and in the case of scenario 1 and 3, a loss of 900 million HUF.

13.7 Stakeholder Workshop

Similar to other projects, like ULYSSES (Dahinden et al. 2000), the final stakeholder workshop was moderated by two experts, one group moderator and one model moderator. The group moderator guided the discussions during the workshop, and the model moderator guided the specific discussions during the computer interaction period. Initially, the model moderator presented the simulation model briefly to clarify the simulation modelling concept, the terminology and how to interpret generated results.

---

[2] 1,000 HUF equals approximately 4.7 USD.

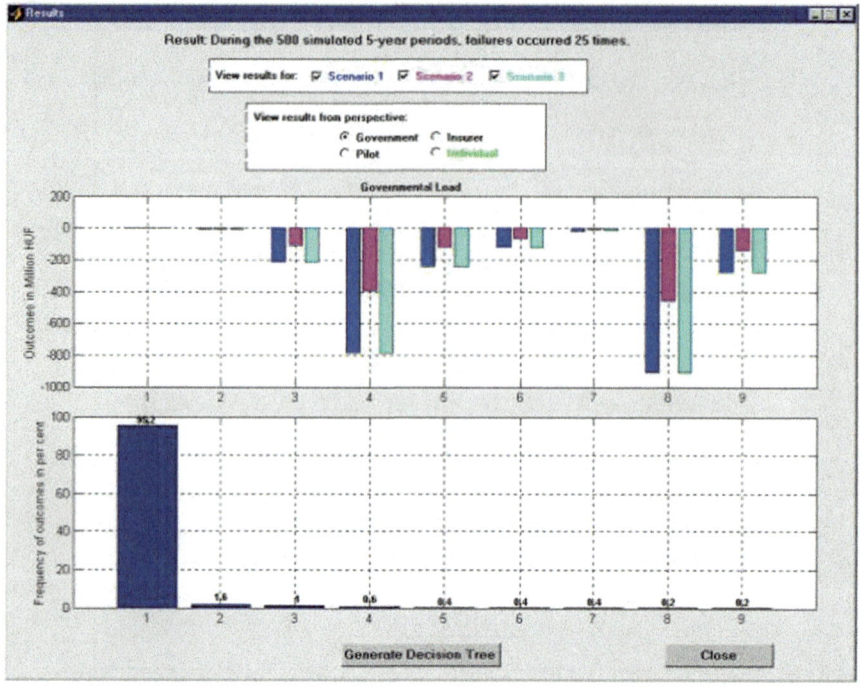

Fig. 13.4 The results window. The *top graph* shows the different types of outcomes resulting from the simulation. The *bottom graph* shows the corresponding frequency of each type of outcome

The stakeholders (representatives from the four stakeholder groups) were divided into three groups depending on which of the three pre-defined policy options they preferred. After extensive model-supported deliberations, the participants reached consensus on a new scenario (Policy Option D), the design of which was made on screen interactively using the Experiment mode:

Consensus Option (D)

- Government compensation only to insured households (50% of losses);
- A private insurance system with

  - bundled or separate policies for all types of natural disaster risks (not modelled),
  - covering approximately 50% of the damage,
  - voluntary, flat-rate premiums (0.01% of property value),
  - no government reinsurance;
  - Government subsidies for poor households up to 100% of premium

After reaching consensus, there was an open discussion of the results and the use of the tool. The participants had a positive attitude towards such a tool for facilitating deliberative participatory processes.

13.8 Conclusions

The model-based stakeholder process described in this chapter shows that an interactive simulation model can aid a deliberative participatory process. A key ingredient to success is early involvement and engagement of the stakeholders; their perspectives and views should be represented in the model in a way that is acceptable to them. This means that the design and implementation of the simulation model must be iterative and performed in a co-operative manner. If the model is developed according to these guidelines and provided with a user interface that is easy to understand and manipulate, it can be used gainfully as a decision tool to support policy discussions. During the workshop (described in Chap. 12 by Linneroth-Bayer et al. in this volume), participants were able to support their arguments with results from the simulations and explore new policy options.

Another important ingredient is rapid model execution. To be of practical use during a workshop, a simulation run must not take more than approximately 1 min, and the simulation outcomes must be presented to the users in a fashion useful for comparisons and discussions. These two factors, combined with an understanding of the model and its limitations, facilitated the process of reaching consensus on a difficult and contentious policy issue.

References

Amendola A, Ermoliev Y, Ermolieva T et al (2000a) A systems approach to modeling catastrophic risk and insurability. Nat Hazards 21(2/3):381–393

Amendola A, Ermoliev Y, Ermolieva T et al (2000b) Earthquake risk management: a case study for an Italian region. In: Proceedings of the second Euroconference on global change and catastrophe risk management: earthquake risks in Europe. IIASA, Laxenburg, Austria, 6–9 July 2000

Cain J, Jinapala K, Makin IW et al (2003) Participatory decision support for agricultural management. A case study from Sri Lanka. Agr Syst 76:457–482

Dahinden U, Querol C, Jäger J et al (2000) Exploring the use of computer models in participatory integrated assessment – experiences and recommendations for further steps. Integr Assess 1:253–266

Ermolieva T (1997) The design of optimal insurance decisions in the presence of catastrophic risks. IIASA Interim report IR-97068, International Institute for Applied Systems Analysis, Laxenburg, Austria

Ermolieva T, Ermoliev Y (2005) Catastrophic risk management: flood and seismic risks case studies. In: Wallace SW, Ziemba WT (eds) Applications of stochastic programming, MPS-SIAM series on optimization. SIAM, Philadelphia

Ermolieva T, Ermoliev Y, Fischer G et al (2003) The role of financial instruments in integrated catastrophic flood management. Multinatl Financ 7(3/4):207–230

Friedman B (2004) Value sensitive design. In: Bainbridge WS (ed) Encyclopedia of human-computer interaction. Berkshire Publishing Group, Great Barrington, pp 769–774

Gregory R, Fischhoff B, McDaniels T (2005) Acceptable input: using decision analysis to guide public policy deliberations. Decis Anal 2(1):4–16

Intergovernmental Panel on Climate Change (IPCC) Working Group III (2000) Emission scenarios: summary for policymakers. Tech rep ISBN: 92-9169-113-5, IPPC (WMO UNEP)

Jakeman AJ, Letcher RA, Norton RA (2006) Ten iterative steps in development and evaluation of environmental models. Environ Model Software 21:602–614

Jiggins JLS, de Zeeuw H (1992) Participatory technology development in practice: process and methods. In: Reijntjes C, Haverkort B, Waters-Bayer A (eds) Farming for the future; an introduction to low-external-input and sustainable agriculture. Macmillan, London

KSH (2000) Major annual figures – regional. Hungarian Statistics Central Office, Budapest

Linnerooth-Bayer J, Amendola A (2000) Global change, catastrophic risk and loss spreading –issues of efficiency and equity. Geneva Pap Risk Insur 25(2):203–219

Linnerooth-Bayer J, Vari A, Thompson M (2006) Floods and fairness in Hungary. In: Verweij M, Thompson M (eds) Clumsy solutions for a complex world: governance politics and plural perceptions. Palgrave Macmillan, Basingstoke/New York

Macintosh A, Whyte A (2006) Evaluating how eParticipation changes local democracy. In: IraniZ, Ghoneim A (eds) Proceedings of the eGovernment workshop 2006. Brunel University, London, eGov06

Mitton L, Sutherland H, Weeks M (2000) Microsimulation modelling for policy analysis: challenges and innovations. Cambridge University Press, Cambridge/New York

Rios Insua D, Kersten GE, Rios J et al (2007) Towards decision support for participatory democracy. Inf Syst E-Bus Manag 6(2):161–191

Stern PC, Fineberg HV (1996) Understanding risk – informing decision in a democratic society. National Academy Press, Washington, DC

Vári A, Linnerooth-Bayer J, Ferencz Z (2003) Stakeholder views on flood risk management in Hungary's Upper Tisza basin. Risk Anal 23:537–627

Vari A, Ereifej L, Ferencz Z (2009) Implementing the EU water framework directive in Hungary: a pilot project in the Upper-Tisza region. Int J Risk Assess Manag 12(1):82–102

VITUKI (1999) Consult Rt. Explanation of detailed methodology for flood damage assessment, Budapest

Walker G (1997) Current developments in catastrophe modelling. In: Britton NR, Oliver J (eds) Financial risk management for natural catastrophes. Griffith University, Brisbane

Chapter 14
A Risk-Based Decision Analytic Approach to Assessing Multi-stakeholder Policy Problems

Mats Danielson and Love Ekenberg

Abstract The design of a public-private flood insurance system is a multi-stakeholder policy problem. The stakeholders include, among others, the public in the high-risk and low-risk areas, the insurance companies and the government. With an understanding of the preferences of the stakeholder groups, decision analysis can be a useful tool in establishing and ranking different policy alternatives. However, the design of a nation-wide insurance system involves handling imprecise information, including estimates of the stakeholders' utilities, outcome probabilities and importance weightings. This chapter describes a general approach to analysing decision situations under risk involving multiple stakeholders. The approach was employed to assess options for designing a public-private flood insurance and reinsurance system in Hungary with a focus on the Tisza river basin. It complements the actual stakeholder process for this same purpose described in previous chapters of this book. The general method of probabilistic, multi-stakeholder analysis extends the use of utility functions for supporting evaluation of imprecise and uncertain information.

Keywords Flood insurance • Multi-stakeholder policy problem • Decision analysis • Upper Tisza river basin

14.1 Introduction

The primary purpose of this chapter is to demonstrate a methodological approach able to cope with multi-stakeholder decision problems in disaster risk management. The methodology was applied to the Tisza river case study described in Chaps. 12 and 13. The upper Tisza river is subject to annual floods, and extreme floods are

M. Danielson • L. Ekenberg (✉)
Department of Computer and Systems Sciences,
Stockholm University, SE-164 40 Stockholm, Kista, Sweden
e-mail: lovek@dsv.su.se

A. Amendola et al. (eds.), *Integrated Catastrophe Risk Modeling: Supporting Policy Processes*, Advances in Natural and Technological Hazards Research 32,
DOI 10.1007/978-94-007-2226-2_14, © Springer Science+Business Media Dordrecht 2013

expected every 10–12%;years (Vári 1999). Financial losses from floods are severe in this region, and costs for compensation to victims and mitigation strategies are increasing. At the time of this study, the government was seeking alternative flood management strategies, where part of the economic responsibility would be transferred from the public to the private sector. This meant finding a balance between social solidarity and private responsibility. Most Hungarians, however, viewed the government as primarily responsible, meaning it should both protect them from flooding and compensate them for flood losses. The government considered this policy as no longer affordable. Moreover, a flood risk management policy would need to consider other views, including those of residents in high – and low-risk areas, insurers, the tourist and other industries, farmers, environmental groups and other NGOs, (non-governmental organizations). Consequently, there was a strong need for reaching a consensus on loss sharing policies that would include stakeholders.

As discussed in Chaps. 12 and 13, the design of a public-private flood insurance system for Hungary presents a significant multi-stakeholder policy problem, and this article focuses on its decision analytical component. In particular we apply a decision tree evaluation method integrated with a common framework for analysing multi-stakeholder decision situations under risk. The background data for the analysis was provided by the Hungarian Academy of Sciences and complemented by stakeholder interviews and a simulation model for investigating the effects of selected policy options for a flood risk management program in the region (see also Brouwers et al. 2004).

Since standard decision analytical methods are not suitable for this multistakeholder problem, we used an interval-based method. Stakeholders rank the relevant attributes of the policy option, such as whether it includes subsidies to low-income households or presents low or high risks to the private insurers. Furthermore, the design of a nation-wide insurance system involves imprecise information, including estimates of the stakeholders' utilities, outcome probabilities, and weights.

Based upon statistical data and interviews (Brouwers and Riabacke 2012, Chap. 13 this book), we demonstrate how implementation of a simulation and decision analytical model can provide insights on the desirability of the selected general policy options for a flood risk management program in the region. The emphasis is on the multi-criteria and multi-stakeholder issues as well as on the high degree of uncertainty in the background data.

14.2   A Decision Theoretical Approach

The advantages and disadvantages of approaches for evaluating imprecisely stated decision problems are discussed elsewhere (see Ekenberg 2000; Ekenberg and Thorbiörnson 2001; Ekenberg et al. 2005; Danielson and Ekenberg 2007). Our selected decision analytic approach as it is applied to the Tisza multi-stakeholder

Fig. 14.1 A decision tree for decisions under risk

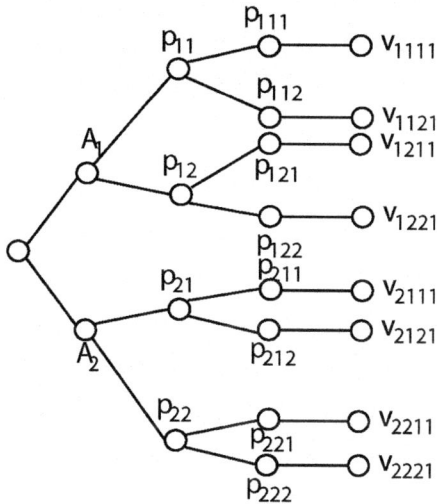

policy problem, i.e., the design of a Hungarian flood insurance system that maximizes stakeholder utilities, is discussed below.

14.2.1 Decision Analysis

Decisions under risk are often represented by decision trees as shown in Fig. 14.1.

The decision tree consists of (1) a root, representing the decision, (2) a set of intermediary (event) nodes, representing uncertain events, and (3) consequence nodes, representing possible final outcomes. Usually, probability distributions are assigned as weights in the event nodes as measures of the uncertainties involved. For an alternative A_i, there is a probability p_{ij} that the accompanying event occurs. This event can have a consequence with a value v_{ijk} assigned to it or a subsequent event. Commonly, the maximization of the expected value is used as an evaluation rule. For instance, in Fig. 14.1, the expected value of alternative A_i is

$$E(A_i) = \sum_{j=1}^{2} p_{ij} \sum_{k=1}^{2} p_{ijk} v_{ijk1} \qquad (14.1)$$

There are several methodologies for analysing multi-criteria, multi-stakeholder problems. However, these methodologies typically do not take adequate account of uncertainty in the decision process. Precise decision information is generally not available in public decision support situations.

The Tisza policy context is characterized by severe uncertain information, which makes it suitable for our structured decision analytic method (Danielson 2004;

Danielson and Ekenberg 2007). There are some approaches that are also interesting candidates for this analysis, such as the generic multi-attribute analysis (GMAA) system (Jiménez et al. 2006) that takes account of uncertainty by allowing value intervals to represent incomplete information about the alternative consequences as well as the decision-maker's preferences. Salo and Hämäläinen (1995) have also developed a set of tools that have been discussed e.g. in (Danielson et al. 2008). An important criterion in choosing the appropriate methodology for the Tisza case is its acceptance by stakeholders.

The method used for evaluating the policy decision problem in the pilot basin was developed for use in large decision problems similar to this case, with several stakeholders and requirements of transparency. It is a multi-stakeholder extension of a method for analyzing decisions containing imprecise information represented as intervals and as qualitative estimates (Danielson and Ekenberg 2007; Ekenberg et al. 2001; Danielson 2004).

In an interval decision analytic method, problem statements are explicitly given, but as intervals or comparisons instead of fixed numbers, and only with the precision the decision makers feel they have evidence for at that moment, e.g., the utility of a consequence is greater or less than the utility of another consequence, or that the probability of an event is within an interval between 30 and 40%. This has a number of advantages. First, the underlying information must be made explicit, and second, the statements can be discussed with (and criticized by) other participants in the decision process. Third, the requisite precision for a decision becomes clearer, including needs for more information. Fourth, arguments for (and against) a specific decision can be derived from the analysis material. Fifth, the decision can be better documented, and the underlying information as well as the reasoning leading up to a decision can be traced afterwards. The decision can even be changed in a controlled way should new information become available at a later stage.

14.3   The Policy Problem Formulation

As described in Chap. 13, two stochastic variables are used to represent flood uncertainties, and the decision problem is based on various scenarios given the stated sets of financial indicators.

Based on this, we have four prototypical scenarios. The first scenario is a private insurance alternative with government subsidies for the poor, and the government acts as reinsurer. This is option B1 in Brouwers and Riabacke (2012).

- No post-disaster government compensation to private flood victims
- Private insurance system characterized by

  – non-mandatory policies,
  – cover for flood, standing water, and all hazards (not modeled),

- insurance purchased separately or bundled with property insurance policies,
- up to 100% cover for damage without deductibles, and
- risk-based premiums;

- Government subsidies for poor households up to 100% of premium;
- Government re-insurance fund financed by tax revenues.

In the second scenario, the government compensates flood failure victims. Private insurance with government subsidies for the poor are used and the government re-insure. The assumptions are the following:

- Government compensation to private flood victims for 50% of their damages
- Private insurance system

  - Private, non-mandatory policies
  - Natural disaster (flood, standing water, earthquake, etc.) insurance can be purchased separately or bundled with property insurance policies
  - Can reach 50% of the damage
  - No risk-based pricing for natural disasters (cross subsidies within system)

- Government subsidies for poor persons to purchase natural disaster insurance (can reach 100%)
- Government acts as re-insurer of last resort. Government re-insurance fund financed by tax revenues
- Contribution to prevention fund by insurance companies.

The third scenario is a variant of the second, where the government compensates flood failure victims, but does not re-insure. The assumptions are the following:

- Government compensation to private flood victims for 50% of their damages – contingent upon the purchase of natural disaster insurance;
- Private insurance system

  - Private, non-mandatory policies;
    Natural disaster (flood, standing water, earthquake, etc.) insurance can be purchased separately or bundled with property insurance policies;
  - Covers 50% of the damage;
  - No risk-based pricing for natural disasters (cross subsidies within system);

- Government subsidies for poor persons to purchase natural disaster insurance (100%);
- No government reinsurance fund;
- Contribution to prevention fund by insurance companies.

In the fourth scenario, the responsibility is partly shifted from the government to the individual property owner. It includes mandatory public insurance for natural disasters with government subsidies for the poor. The assumptions are the following:

- Public insurance system administered by private insurance companies

  - Public, mandatory policies;
  - Only for natural disasters (flood, standing water, earthquake, etc.);
  - Can reach 100% of the damage;
  - No risk-based pricing (cross subsidies within system);

- Government subsidies for poor persons to purchase natural disaster insurance (can reach 100%);
- Government establishes catastrophe fund to compensate natural disaster victims and to assume all risks (risk of diverting catastrophe fund).

14.4   Analysing the Policy Scenarios

The computer tool DecideIT was employed (Danielson et al. 2003) for the evaluations. The tool enables handling the information and making evaluations in an automated way. The values from the Tisza investigations were entered into the tool from the simulations similar to Brouwers and Riabacke (2012).

14.4.1   Representation of the Decision Problem

The decision tree is generated from the four policy scenarios, which are represented as alternative branches. The final outcomes are divided into the three stakeholder categories: Government, Insurance companies, and Pilot basin (i.e. municipalities). Figure 14.2 shows a simplified tree used for option analyses. Figure 14.3 shows a part of the actual expansion of one of the final nodes in Fig. 14.2.

In the absence of an actual and precise uncertainty measure, the simulations serve as a basis for a more elaborate sensitivity analysis that will consider both probabilities for floods and estimates of losses. The frequency of floods and levee failures used in the background simulations is based on historical data. In general, simulations of this type are dependent on a large number of input data, are, as such, are sensitive to various types of errors. For instance, the simulations do not reflect the flood frequency and magnitude (peak) increase during recent years. This may be a result of the changes in forestation, urbanization, asphalting, and other land uses or climate change (IPCC 2001). In the analyses below, ranges of values have been used instead of the precise values from the simulations. The ranges are intervals centred on the simulation values as mid-points.

Furthermore, since no structural mitigation measures are taken into account, the decision problem is a zero-sum game, i.e., the problem is to find how the relative importance of the stakeholders should be taken into account. This can be done by assigning their preferences different weights. Naturally, there are severe difficulties in assigning precise numbers for these importance weights, and it is more natural to study the effects from weaker qualitative statements such as *"the importance of*

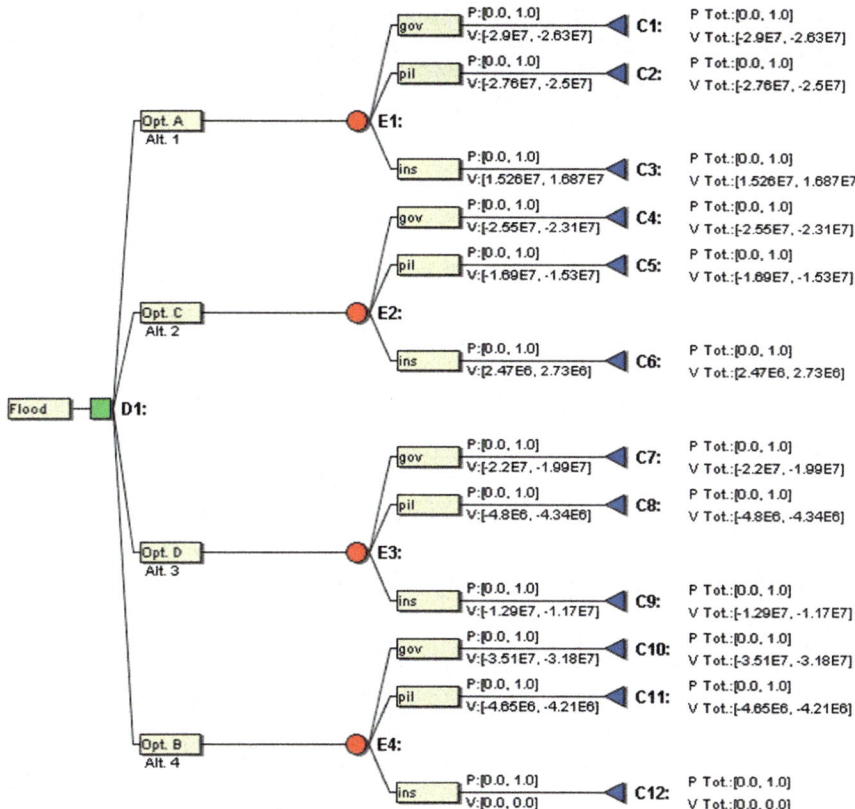

Fig. 14.2 The simplified decision tree

stakeholder 1 is greater than the importance of stakeholder 2", etc. The effects of altering the feasible space of the importance weights are then easily analysed in the tool.

14.4.2   Evaluation of the Policy Options

Using the simulation results from Brouwers and Riabacke (2012) in Chap. 13 of this book, the analyses incorporate the stakeholder preferences, costs of the different mitigation measures as well as costs for damages and property losses, and probabilities of the different flood scenarios. In the Tisza case, three stakeholder categories were modelled and the evaluation procedure is described below.

The primary evaluation rule for the decision trees is based on the generalized expected values of the scenarios, taking all probabilities, values and (in the final analysis) criteria weights into account. Since neither probabilities nor values are

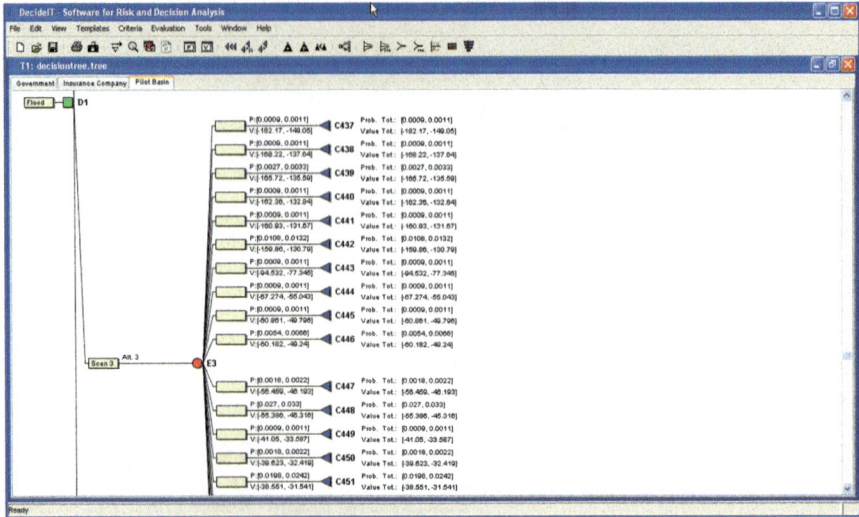

Fig. 14.3 Small part of the decision tree as viewed in the tool

fixed numbers, the calculation of the expected value yields multi-linear objective functions as in Eq. 14.1 above. Maximization of such non-linear expressions are computationally demanding problems to solve for an interactive tool in the general case, using techniques from the area of non-linear programming. In, e.g., Danielson and Ekenberg (2007) there are discussions about computational procedures to reduce non-linear problems to systems with linear objective functions, solvable with ordinary linear programming methods. Equation 14.1 is then evaluated with respect to the outcomes of the stakeholders and a total value range for the alternatives is then obtained by weighting these results with respect to the stakeholders' assigned relative importance. Since it is practically impossible to assert a meaningful quantitative (numeric) stakeholder weight (e.g. $w_1 = 0.427$), qualitative weights are used instead in the analysis (e.g. $w_1 > w_2$). This is easier to understand for the participants in the process.

How is it now possible to compare the scenarios? For a regular decision tree with decision nodes, event nodes, and consequence nodes, there is one probability constraint, **P,** and one value constraint, **V**. For a multi-criteria model, as in the case here, there is also a weight constraint. When the probabilistic decision trees are evaluated under some stakeholder preferences, the probability variables, value variables and related constraints are assigned to the alternatives of the model. For a decision tree, variables and constraints are assigned to the consequence nodes.

To further aid in the modelling of the problem, the *orthogonal hull* concept indicates to the decision-maker which parts of the statements are consistent with the available information given. The decision information can be considered as constraints in the space formed by all decision variables. The (orthogonal) hull is then the projection of the constrained spaces onto each variable axis, and can thus be seen as the meaningful interval boundaries (Danielson and Ekenberg 2007). The same type of input is used for values, probabilities, and weights, although the normalization constraints $\Sigma\ p_i = 1$ (for probabilities) and $\Sigma\ w_j = 1$ (for weights) must not be violated. All input into the model is subject to consistency checks performed by the tool.

For each variable, there is also a *focal point*, which may be viewed as the 'most likely' or 'best representative' value for that variable. Hence, a focal point is a unique solution vector whose components for each dimension (variable) lies within the orthogonal hull. Given this, we calculate the *strength* of alternatives as a means for further discriminating the alternatives. The strength δ_{ij} simply denotes the difference in expected value between two scenarios (alternatives) A_i and A_j, i.e. the expression $\mathbf{E}(A_i)-\mathbf{E}(A_m)$. For multi-criteria models, the expected value for each criterion is aggregated into a weighted sum of expected values for the entire decision problem. By denoting the expected value of an alternative A_i with respect to the kth stakeholder with ${}^k\mathbf{E}(A_i)$, this leads to an expression for the weighted strength

$$\sum_k w_k \left({}^k E(A_i) - {}^k E(A_j) \right)$$

In its most basic form (one-level decision tree), ${}^k\mathbf{E}(A_i)$ is reduced to

$$\sum_k p_{ik} \cdot v_{ik} - \sum_l p_{jl} \cdot v_{jl}$$

over all consequences belonging to alternative A_i and A_j respectively, such that p_{il} denotes the probability of the lth consequence, possibly occurring when choosing scenario A_i. Details on how this is computationally handled in the evaluation are found in (Larsson et al. 2005; Danielson and Ekenberg 2007). Hence, in the tool, probabilistic decision trees may be used alone for single-objective decision problems and can also be "connected" at any time to a criterion leaf-node in the criteria tree as long as the initial alternatives in the probabilistic decision trees map one-to-one onto the alternative set in the multi-criteria tree.

In the evaluation, the alternatives are pair-wise compared and a ranking is induced. The strengths of each alternative compared to all the others can then be compared using the tool. This results in graphs showing the maximum strengths of the alternatives.

Figure 14.4 shows the result of asserting government as the most important stakeholder. The importance weights were set accordingly (i.e. w(Gov) > w(Mun) and w(Mun) > w(Ins) for a weight function w(·) that sums to one) and the probabilities and costs were provided from the simulations. Note that no explicit numerical weights had to be supplied. The x-axis shows the base cut, which is a sensitivity analysis zooming in on central parts of the intervals. See below for a more detailed discussion on the concept. The y-axis shows the difference in

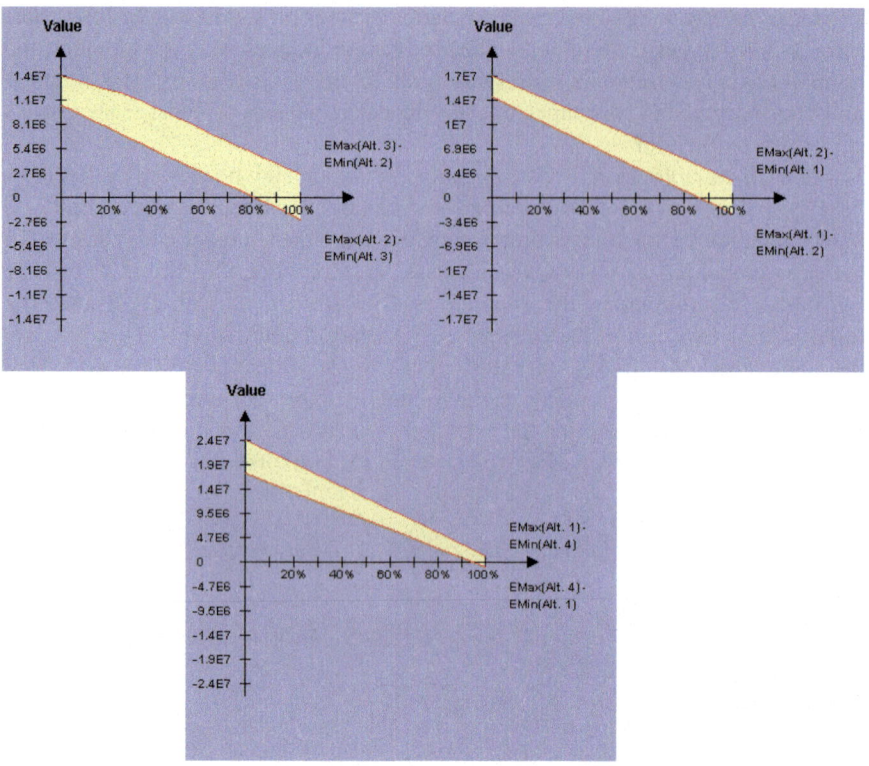

Fig. 14.4 Government is considered to be the most important category

strength.[1] Figure 14.4 shows graphs, representing the strengths of the alternatives. at various base cuts. Thus, the upper graph represents the most preferred alternative. Thus, in Fig. 14.4, the ranking of the alternatives is (from most to least preferred): Alt. 3 (Refined Alt. 2), Alt. 2 (Mixed Insurance), Alt. 1 (Individualistic), and Alt. 4 (Public). It should be emphasized that this does not mean that it is impossible for, e.g., Alt. 2 to be more favourable than Alt. 3. As long as the graph of Alt. 2 is above the x-axis, there is such a possibility. However, the likelihood that Alt. 3 is the alternative to prefer is much higher, so this should be chosen if no other information is available. A natural way to handle the inherent imprecision is to consider values near the boundaries of the intervals as being less reliable than more central ones. If the strength is evaluated on a sequence of ever-smaller sub-bases, a good appreciation of the strength's dependency on boundary values can be obtained. This is taken into account by cutting off the dominated regions indirectly. This is called *cutting the bases*, and the amount of cutting is indicated as a percentage, which can range from 0 to 100%.

---

[1] The scales are from the lowest value to the highest value on the y-axes.

This can be seen as the x-axis in Fig. 14.4, which shows progressively larger cuts. For a 100% cut, the bases are transformed into single points (focal points), and the evaluation becomes the calculation of the ordinary expected value. It is possible to regard the hull cut as an automated kind of sensitivity analysis. Since the belief in peripheral values is somewhat less, the interpretation of the cut is to zoom in on more believable values that are more centrally located. Or conversely, to zoom out from the focal points, adding uncertainty as the zooming out progresses (leftwards in the figures). Thus, this kind of contraction along the x-axis is a sensitivity analysis procedure, in which all intervals are compressed in a controlled way towards the focal point of the multi-dimensional space that all consequences span. Using cut levels as in Fig. 14.4, it can be seen that δ_{23}, δ_{12} and δ_{41} are strictly less that 0 at cut levels 80, 85, and 95%, respectively. This means that in a quite substantial volume around the focal point, Alt. 3 is definitely better than Alt. 2, etc., and consequently, that there is no possibility for the converse to hold. Thus, the above ranking is fairly stable under this kind of sensitivity analyses.

More formally, for comparing alternatives A_i and A_j, the upper line is $\max(\delta_{ij})$ and the lower is $-\min(\delta_{ij})$, i.e. the lower line is reversed to facilitate an easier comparison. Thus, one can see from which cut level an alternative dominates another. As the cut progresses, one of the alternatives eventually dominates strongly, i.e. there are no variable assignments yielding $\max(\delta_{ij}) > 0$. The interpretation of the figures will be further described below.

Figure 14.5 shows the analysis when the difference of the weight of the government is greater than the weight of the municipalities plus a constant of 0.3, i.e., the government is perceived to be of even greater importance. As in the previous analysis, both these weights are greater than the weight of the insurance companies. It can be seen from the figure that the preference order is still the same as above, but that the relative difference between these have increased. The x-axis shows the cut in per cent ranging from 0 to 100. The y-axis is the expected value difference δ_{ij} for the pairs. Using the same kind of analysis as above, it can be seen from Fig. 14.5 that the ranking of the alternatives this time is the same.

Figure 14.6 shows the result of asserting the municipalities as being the most important stakeholders (i.e. w(Mun) > w(Ins) and w(Mun) > w(Gov)). The ranking of the alternatives is (from most to least preferred): Alt. 3, Alt. 4, Alt. 2, and Alt. 1.

We will here refrain from any recommendations concerning the best overall choice, but note that during the discussions with the stakeholders, it seemed to be a common view that at least the insurance companies should not be considered the most important. If also it is assumed that the government and the municipalities are considered as more important than the insurance companies (i.e. w(Mun) > w(Ins) and w(Gov) > w(Ins)), we receive the result that the final alternative from the stakeholder workshop is the preferred one. Now, one might argue that Alt. 4 ought to be slightly modified to better fit also the perspective of the insurance companies. In this way, information from the analyses can be fed back into the decision process, yielding modified alternatives that in the end would be more acceptable to a majority of the stakeholders. Such feedback loops are typical when working with a method allowing imprecise data.

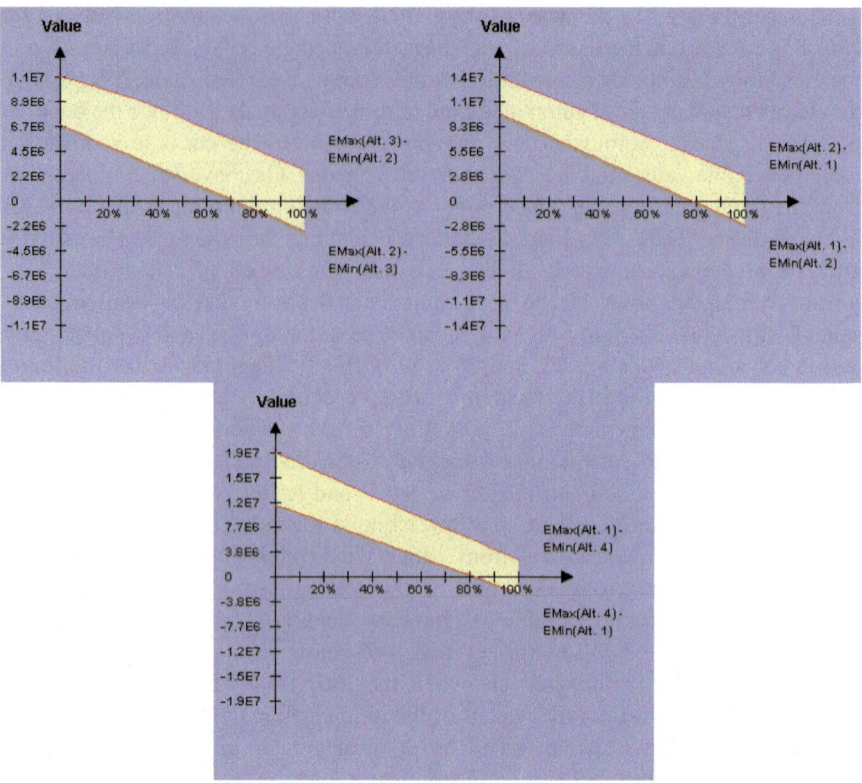

Fig. 14.5 Weight of government much greater than municipalities

In conclusion, this brief analysis shows the stability of the results. Importantly, it is not necessary to assign explicit weights to the stakeholders (which could be difficult or even controversial) to obtain this result. It is sufficient to give general and comparatively weak preferences and still obtain confidence in the result.

It should be emphasized that the figures above are intended to explain the basics of the method and do not present conclusive analyses of the Tisza case. The analysis can (and should) be extended by, at least, further sensitivity analysis, analysis of critical factors, and settings of security levels (Ekenberg 2000).

14.5   Summary and Concluding Remarks

We have discussed a decision analytical method for investigating the effects of different policy options for a flood risk management program in the Tisza region. This multi-stakeholder problem could not be solved using standard approaches, and we have used an interval approach, called EDM, for the decision analytical part.

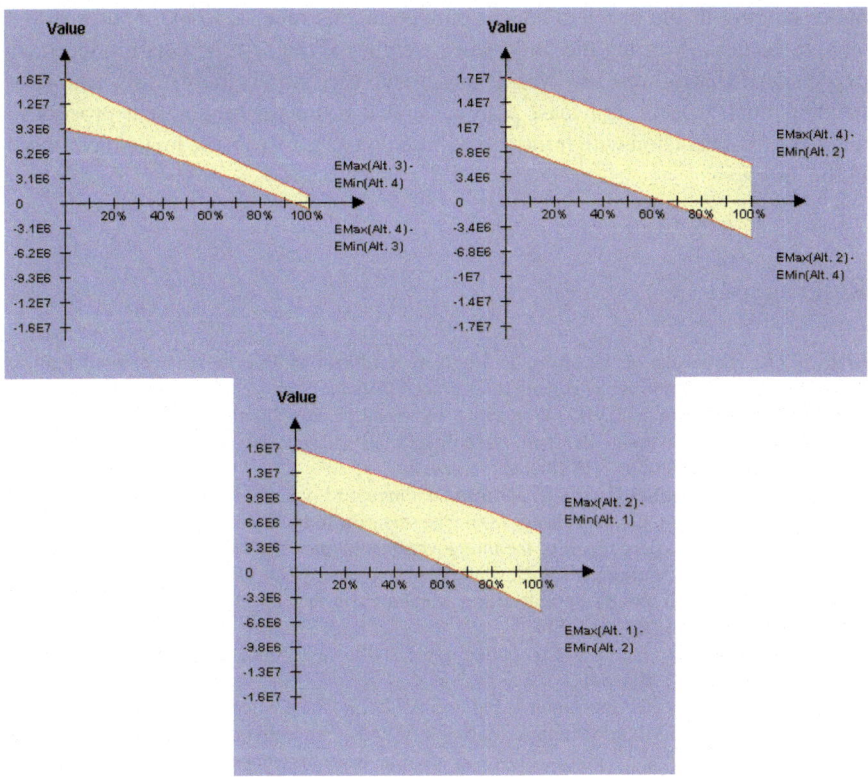

Fig. 14.6 Municipalities is considered to be the most important category

We have focused on the analysis of four alternative flood management policy strategies, and used computational decision analysis to investigate the strategies. We have explored the effects of imposing the strategies for the purpose of illuminating significant effects of adopting different insurance policies. The main focus has been on insurance schemes in combination with the level of government compensation.

The analyses of the different policy strategies have been based on a simulation model where expected flood failures as well as geographical, hydrological, social, and institutional data have been taken into account. The generated results were thereafter transposed to decision trees under three stakeholder perspectives. Taking the simulation results into account, the scenarios have been analysed with the decision tool DecideIT for evaluating the policy options under the various costs, importance weights, and probabilities involved. No explicit importance weights for the stakeholders were necessary, but only an ordering among them.

Note that we do not present all analyses performed, since the primary purpose is to demonstrate a methodological approach to the multi-stakeholder decision problem. It is important to note that the end result of the analysis is a thorough understanding of the problem and recommendations relative to the input material and preferences. Varying the importance weights of the stakeholders produced, as expected, differing results. Needless to say, the issues involve several non-mathematical aspects, not least political and it is up to the political process to make the final decision with information based on results from the method proposed in this article.

References

Brouwers L, Danielson M, Ekenberg L, Hansson K (2004) Multi-criteria decision-making of policy strategies with public-private re-insurance systems. Risk Decis Policy 9(1):23–45

Brouwers L, Riabacke M (2012) Consensus by simulation: a flood model for participatory policymaking. In: Amendola et al. (eds) Integrated catastrophe risk modeling: supporting policy processes. Springer, Dordrecht, pp xx–xx

IPCC (2001) In: Houghton JT, Ding Y, Griggs DJ, Noguer M, van der Linden PJ, Dai X, Maskell K, Johnson CA (eds) Climate change 2001: the scientific basis. Contribution of working group I to the third assessment report of the intergovernmental panel on climate change. Cambridge University Press, Cambridge, UK/New York, 881 pp

Danielson M (2004) Handling imperfect user statements in real-life decision analysis. Int J Inf Technol Decis Mak 3(3):513–534

Danielson M, Ekenberg L (2007) Computing upper and lower bounds in interval decision trees. Eur J Oper Res 181:808–816

Danielson M, Ekenberg L, Johansson J, Larsson A (2003) The DecideIT decision tool, ISIPTA '03. In: Bernard J-M, Seidenfeld T, Zaffalon M (eds) Proceedings of the 3rd international symposium on imprecise probabilities and their applications, proceedings in informatics 18, Carleton Scientific, pp 204–217

Danielson M, Ekenberg L, Ekengren A, Hökby T, Lidén J (2008) A process for participatory democracy in electronic government. J Multi-Criteria Decis Anal 15(1–2):15–30

Ekenberg L (2000) Risk constraints in agent based decisions. In: Kent A, Williams JG (eds) Encyclopaedia of computer science and technology 23(48):263–280, Marcel Dekker Inc.

Ekenberg L, Thorbiörnson J (2001) Second-order decision analysis. Int J Uncertain Fuzziness Knowl-Based Syst 9(1):13–38

Ekenberg L, Boman M, Linnerooth-Bayer J (2001) General risk constraints. J Risk Res 4(1):31–47

Ekenberg L, Thorbiörnson J, Baidya T (2005) Value differences using second order distributions. Int J Approx Reason 38(1):81–97

Jiménez A, Rios-Insua S, Mateos A (2006) A generic multi-attribute analysis system. Comput Oper Res 33(4):1081–1101

Larsson A, Johansson J, Ekenberg L, Danielson M (2005) Decision analysis with multiple objectives in a framework for evaluating imprecision international. J Uncertain Fuzziness Knowl-Based Syst 13(5):495–510

Salo AA, Hämäläinen RP (1995) Preference programming through approximate ratio. Comp Eur J Oper Res 82:458–475

Vári A (1999) Flood control development in Hungary: public awareness report. Prepared for VITUKI consult Rt Budapest: Hungarian Academy of Sciences Institute of Sociology

Chapter 15
Optimizing Public Private Risk Transfer Systems for Flood Risk Management in the Upper Tisza Region

Yuri Ermoliev, Tatiana Ermolieva, and Istvan Galambos

Abstract This chapter summarizes studies on the development of a financial risk management model for floods in the Upper Tisza river region, Hungary. We focus on the evaluation of a multi-pillar flood loss-spreading program involving partial compensation to flood victims by the central government, the pooling of risks through a mandatory public-private insurance on the basis of location-specific exposures, and a contingent ex-ante credit to reinsure the pool's liabilities. Policy analysis is guided by GIS-based catastrophe models and stochastic optimization methods with respect to location-specific risk exposures. We use economically sound risk indicators leading to convex stochastic optimization problems strongly connected with non-convex insolvency constraint and Conditional Value-at-Risk (CVaR).

Keywords Flood risk • Catastrophe modeling • Natural risk insurance • Stochastic optimization • Contingent credit • CVaR

Y. Ermoliev
International Institute for Applied Systems Analysis (IIASA),
Schlossplatz 1, A-2361 Laxenburg, Austria

T. Ermolieva (✉)
Ecosystems, Services and Management (ESM) Program, International Institute for Applied Systems Analysis (IIASA), Schlossplatz 1, A-2361 Laxenburg, Austria
e-mail: ermol@iiasa.ac.at

I. Galambos
Center for Water Systems, University of Exeter, Exeter, UK

A. Amendola et al. (eds.), *Integrated Catastrophe Risk Modeling: Supporting Policy Processes*, Advances in Natural and Technological Hazards Research 32,
DOI 10.1007/978-94-007-2226-2_15, © Springer Science+Business Media Dordrecht 2013

15.1 Introduction[1]

Inadequate land use practices, deforestation, clustering of people and capital in hazard-prone areas, phenomena aggravated by lack of knowledge or ignorance of the risks, are among the main reasons for the increasing dimensions of natural disasters. Indeed, global scale analysis of natural disaster hotspots (Dilley et al. 2005) shows that damages from natural catastrophes have been rising just because of economic growth in hazard-prone areas.

This alarming tendency calls for new integrated approaches to catastrophic risk management in hazard prone areas. The existing approaches often ignore rare disasters of high consequences and their complex spatio-temporal heterogeneities, effective management of which requires a variety of interdependent strategies. New systemic approaches were developed for Hungary, where 23% of the country is endangered by riverine floods. Only the Netherlands has a similar degree of risk, with 20% of the country under sea level. In Hungary, as described in Chaps. 11 and 13 of this book, the losses from floods and other natural disasters have been mainly absorbed by the victims and governments (see also Kunreuther and Linnerooth-Bayer 2000). With increasing losses from floods, many governments are concerned with escalating costs for flood prevention, flood response, compensation to victims, and public infrastructure repair. They may attempt to increase the responsibility of individuals and local administrations and to promote risk sharing provisions for flood risks and losses. Enforcement of loss-reduction and loss-spreading measures is possible only after analysis of location-specific potential losses, their mutual interdependency and sensitivity to new risk management strategies. Historical data, even when available, are not very useful in a changing environment. Purely adaptive learning-by-doing type of approaches may be very expensive and dangerous. The availability of models enabling the simulation of probable catastrophes for designing preparedness programs becomes a key task. For this purpose, a new comprehensive model has been developed as part of a joint IIASA-Sweden-Hungary project on Flood Risk Management in the Upper Tisza Basin (Project Proposal 2000).

In most countries (see Froot 1997), losses from disasters . . . "are paid ex-post by some combination of insurers and reinsurers (and their investors), insured, state and federal agencies and taxpayers, with only some of these payments being explicitly arranged ex-ante. This introduces considerable uncertainty about burden sharing into the system, with no particular presumption that the outcome will be fair. The result is incentives for players to shift burdens towards others, from the homeowner who builds on exposed coastline, to insurers who write risks that appear highly profitable in the absence of a large event. But most importantly, bad or inefficient

---

[1] This chapter is based on the paper "The role of financial instruments in integrated catastrophic flood management" by Ermolieva T, Ermoliev Y, Fischer G, Galambos I, *Multinational Finance Journal* 7(3&4): 207–230, 2003. The paper is reprinted and modified with permission of the publisher.

risk sharing raises the cost of capital for companies and requires returns for households, reducing the amount of profitable investments and the rate of growth of the economy." Ex-ante mechanisms to fund the costs of recovery and, in particular, the establishment of a multi-pillar flood loss-sharing program, are especially important.

In our analysis we assume that for the first pillar the Hungarian government would provide compensation of a limited amount to all households that suffer losses from flooding. As the second pillar, a special regional fund would be established through a mandatory flood insurance program on the basis of location-specific risk exposures. It is assumed that the governmental financial aid is regulated through this fund. As a third pillar, a contingent credit may also be available to provide an additional injection of capital to stabilize the system. In the latter case, the lender charges a fee that the borrower (in our case, the fund) pays as long as the trigger event does not occur. If the event does occur, the borrower rapidly receives the fund.[2]

The analysis of possible gains and losses from different arrangements of the program is a multi-disciplinary task, which needs to take account of the frequency and intensity of hazards, the stock of capital at risk, its structural characteristics, and different dimension of vulnerability (e.g. engineering, financial). These efforts require the development of comprehensive catastrophe models (Walker 1997). Section 15.2 discusses the main features of a GIS-based catastrophe model developed for the Upper Tisza pilot region that, in the scarcity of historical data, simulates samples of potential losses dependent on location-specific infrastructure, property values, land use practices, etc.

Traditional insurance and finance quantify extreme events relying on available loss estimates (Embrechts et al. 2000). Catastrophes produce interdependent direct and indirect losses which may be unlike anything that has been experienced in the past. The catastrophe model simulating such losses may deal with multivariate dependent distributions of extreme values, i.e., with the case which is not sufficiently addressed within the conventional extreme value theory. The existing catastrophe models would open up the possibility for an "if-then" scenario analysis but such approach suffers from the shortcomings discussed in Sect. 15.3. Those drawbacks are overcome by the Adaptive Monte Carlo (AMC) optimization procedure proposed by Ermolieva et al. (1997) and Ermolieva (1997).

Section 15.4 describes the spatial and dynamic stochastic optimization model developed for the design of public multi-pillar flood loss spreading programs. The model emphasizes the cooperation of various agents in dealing with catastrophes. Sound catastrophic risk management policies, especially for small economies with limited risk absorption capacity, cannot be accomplished without pooling of risk exposures (see discussion in Pollner 2000; Amendola et al. 2000a, b; Cummins and Doherty 1996). The proposed model involves pooling risks through mandatory

---

[2] The advantages of this financial arrangement in contrast to catastrophic bonds are discussed, e.g., in Pollner (2000).

flood insurance based on location-specific exposures, partial compensation to the flood victims by the central government, and a contingent credit to the insurance pool. Definitely, this program encourages accumulation of own regional capitals to better "buffer" the volatility of international reinsurance markets. In order to stabilize the program we use economically sound risk indicators such as expected overpayments by "individuals" and expected shortfall of the mandatory insurance (similar to our analysis of seismic risk programs in Amendola et al. 2000a, b, Ermolieva et al. 1997. See also Chap. 3 of this book). We simulate the stochastic system until the first flood when catastrophic losses exceed specified level (the time of this event is defined as "stopping time" event). In this way we orient the analysis towards the most destructive scenarios.

The explicit introduction of ex-post borrowing as a measure against insolvency enables us to approximate the insolvency constraint by a convex optimization problem, whereas the use of the contingent credit leads to the Conditional-Value-at-Risk (CVaR) type of risk measures. Section 15.5 specifies this model further. Numerical experiments indicate a strong dependence of demand for contingent credit on the composition of other risk management measures. The importance of such an integrated analysis was emphasized in Ermoliev et al. (2000a, b), Kleindorfer and Kunreuther (1999), Mayers and Smith (1983).

15.2   A Catastrophe Model

The aim of catastrophe models (Walker 1997) is to generate potential samples of mutually dependent losses for a given vector of policy variables. For example, when there is a lack of historical data or the area undergoes substantial transformations, models can estimate distributions of losses and gains for different locations, households, insurers, and governments. This is critically important in the case of rare events or new policies that have never been implemented in practice. The catastrophe model developed for the pilot region of the Upper Tisza river consists of five submodels (modules): the "River" module, the "Inundation" module, the "Vulnerability" module, the "Multi-Agent Accounting System", and the "Variability" module (see Fig. 15.1).

The River module calculates the volume of water discharged into the study region from the different river sections for given heights of dikes and associated scenarios of their failures or removals, and given rainfall patterns. The latter are modeled by upstream discharge curves. Thus, formally, the River module maps an upstream discharge curve into the volume of water released to the region from various sections. The underlying submodel is able to estimate the discharged volume of the water into the region under different conditions, for example, if the rain patterns change, if the dikes are heightened, or if they are strengthened or removed.

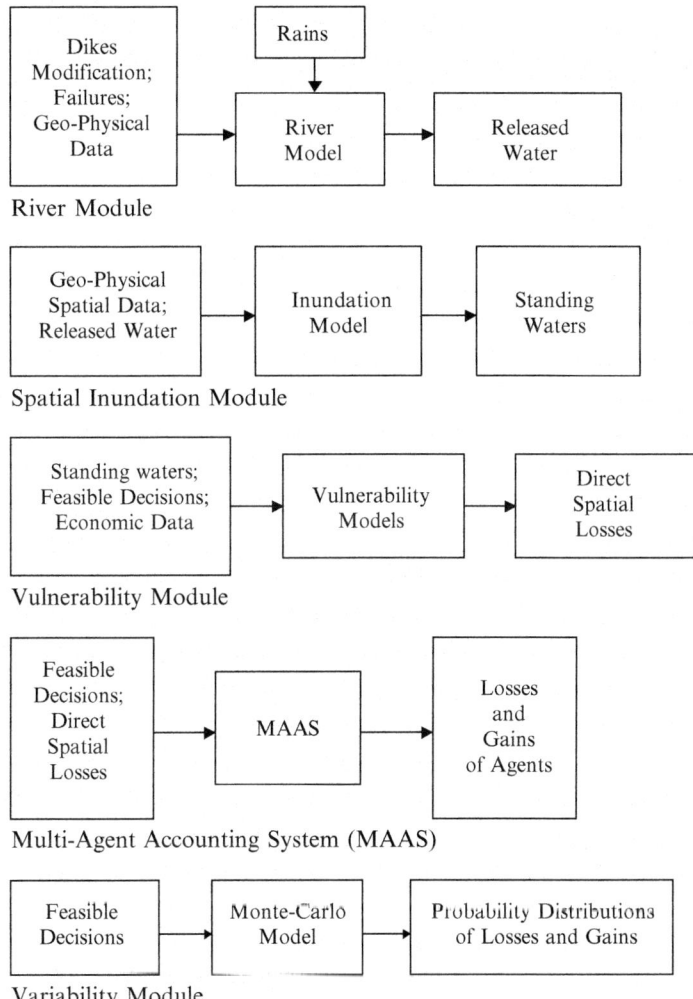

Fig. 15.1 Modules of the catastrophe model

The next module is the spatial GIS-based Inundation submodel. The pilot region has been modeled by a 1,500 × 1,500 grids. This module maps water released from the river into levels of standing water.

The Vulnerability module translates spatial patterns of released water into economic losses. This module calculates direct losses and may include possible cascading effects, such as floods causing landslides and their consequences. It may also include loss reduction measures, e.g., new land-use modifications and flood preparedness measures, simulating thus changes in economic losses induced by risk reduction measures.

The Multi-Agent Accounting System module maps spatial economic losses into gains and losses of agents. These agents are the central government, a mandatory catastrophe insurance pool, an investor, and "individuals".

Given sufficient data, the above-mentioned submodels can generate scenarios of losses and gains at different locations for specific scenarios of failures, precipitation, risk reduction measures and risk spreading schemes. But there are significant uncertainties and a large variability in these losses and gains. A 50-year flood may occur in 5 days or in 70 years. Insurers are especially concerned about variability since they may not have the capacity to cover large losses if they do not accumulate sufficient premiums into their risk reserves before the occurrence of the catastrophic event. In an attempt to maintain their solvency, they may charge higher premiums, which may result in overpayments by the insured. Alternatively, insurers may undercharge contracts if the analysis is inadequate. Uncertainties also may result from adopted loss-reduction measures. For example, increasing the height of a dike may decrease the number of smaller floods and create an illusion of safety attracting more properties towards the region. However, if it fails, it may cause more damages in comparison to a dike without modification. The Variability module, a Monte Carlo model, transforms spatial scenarios of losses and gains among agents into probability distributions. For example, it derives histograms of direct losses at a given location or sub-region. It also calculates histograms of overpayments and underpayments for different agents (see Sect. 15.5).

15.3 Adaptive Monte Carlo Optimization

The catastrophe model opens up the possibility for "if – then" analysis, which allows the analyst to evaluate a finite number of policy alternatives. However, this analysis may run quickly into an extremely large number of possible combinations resulting in unacceptable computational time. For example, when studying a region subdivided into 10 locations (e.g. municipalities) with 10 alternative location-specific risk-related policies (insurance coverage or location specific premiums), the number of alternatives to investigate equals 10^{10}. The analytical intractability of stochastic catastrophe models often precludes the use of standard optimization methods, e.g., genetic algorithms. Stochastic optimization methods (Ermoliev and Wets 1988) are able to evaluate desirable policy measures without the evaluation of all possible options. In particular, the so-called Adaptive Monte Carlo Optimization (Ermolieva 1997; Ermoliev et al. 2000a, b). "Adaptive Monte Carlo" (Pugh 1966) means a technique that makes on-line use of sampling information to sequentially improve the efficiency of the sampling itself. The Adaptive Monte Carlo Optimization model for the Upper Tisza region consists of three interacting blocks: Feasible Decisions, the Monte Carlo Catastrophe Model, and Indicators (Fig. 15.2).

The block Feasible Decisions represents all feasible policies for coping with floods. In general, they may include feasible heights of dikes, insurance coverage, land use modifications, etc. These variables affect performance indicators such as

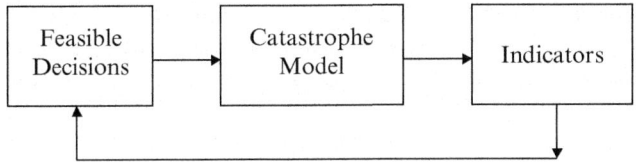

Fig. 15.2 The adaptive Monte Carlo optimization model

profits of insurers, underpayments or overpayments by the insured, costs, insolvency and stability indicators.

The essential feature is the feed-back mechanism updating decisions towards specific goals. Losses are simulated by the catastrophe model, causing an iterative revision of the decision variables after each simulation run. In a sense, the Adaptive Monte Carlo optimization simulates in a remarkably simple and evolutionary manner the learning and adaptation process on the basis of the simulated reversible history of catastrophic events. This technique is unavoidable when the outcomes of the catastrophe model do not have a well-defined analytical structure.

15.4   The Stochastic Optimization Model

In the model for the Upper Tisza region we use approaches similar to those adopted in Amendola et al. 2000a, b, Emolieva et al. 1997, Ermoliev et al. 2000a, b for seismic risks (see also Chap. 3). The main idea is based on subdividing the study region into cells $j = 1, 2, \ldots, m$. These cells may correspond to a collection of households at a certain site, a collection of grids (zones) with a similar land-use structure, or an administrative district or a grid characterizing industrial activities for instance, with a segment of a gas pipeline. The choice of cells provides a desirable representation of losses. In our case, the cells consist of the value of the physical structures. Simulated catastrophic events affect at random different cells at time t and produce mutually dependent losses L_j^t. at time t

If $x = (x_1, x_2, \ldots, x_n)$ is the vector of the decision variables, then losses L_j^t to a cell j at time t are transformed into $L_j^t(x)$. In the case of the Tisza river, for example, we can think of $L_j^t(x)$ as L_j^t being affected by the decisions of the insurance to cover losses from a layer $[x_{j1}, x_{j2}]$ at a cell j in the case of a flood disaster at time t:

$$L_j^t(x) = L_j^t - \max\left\{x_{j1}, \min\left[x_{j2}, L_j^t\right]\right\} + x_{j1} + \pi_j^t \qquad (15.1)$$

where $\max\left\{x_{j1}, \min\left[x_{j2}, L_j^t\right]\right\} - x_{j1}$ are retained by insurance losses, and π_j^t is a premium function. The variable x_{j1} defines the deductible part ("trigger") of the contract and x_{j2} defines its "cap".

In the most general case, vector x comprises decision variables of different agents, including governmental decisions, such as the height of a new dike or a public compensation scheme defined by a fraction of total losses $\sum_{j=1}^{m} L_j^t$. The insurance decisions concern premiums paid by individuals and the payments of claims in the case of catastrophe. There are complex interdependencies among these decisions, which call for the cooperation of agents. For example, the partial compensation of catastrophe losses by the government enforces decisions on loss reductions by individuals and, hence, increases the insurability of risks, and helps the insurance to avoid insolvency. On the other hand, insurance combined with individual and governmental risk-reduction measures can reduce losses, compensations and government debt and stabilize the economic growth of the region and the wealth of individuals.

Let us now turn to consider a potential insurance system for Hungary and introduce some important indicators. In the following, for simplicity of notation, we do not consider the most general situation, e.g., we consider only proportional compensation by the government, proportional insurance coverage and we do not use discount factors.

In this application the system is modeled until a first catastrophic flood, which occurs within a given time horizon (stopping time). For the Upper Tisza region this event is associated with the break of a dike that may occur after a 100-year, 150- or 1,000-year flood. Inputs to the simulation model are upstream discharge curves and probabilities of dike failures. The timing of the first catastrophic flood significantly affects the accumulation of risk reserves by the insurance, and total payments of individuals; for example, a 100-year flood with the break of a dike may occur in 2 years.

Let τ be a random (stopping) time to a first catastrophe within a time interval $[0, T]$, where T is some planning horizon, say, of 10 or 50 years. If no catastrophe occurs, then $\tau = T$. Since τ is associated with the break of a dike, the probability distribution of τ is, in general, affected by some components of vector x, e.g., by decisions on dike modifications, land use changes, building reservoirs, etc. In this chapter we discuss only the case when τ does not depend on x.

Let L_j^τ be random losses at location j at time $t = \tau$. In our analysis we evaluate the capacity of the catastrophe insurance in the upper Tisza region only with respect to financial loss-spreading decisions. Let us use a special notation for their components such as π_j, ϕ_j, v, q, y. If π_j is the premium rate paid by location j to the mandatory insurance, then the accumulated mutual catastrophe fund at time τ together with the proportional compensation $v \sum L_j^\tau$ by the government is equal to $\tau \sum_j \pi_j + v \sum_j L_j^\tau - \sum_j \phi_j L_j^\tau$, where $0 \leq \phi_j \leq 1$, is the insurance coverage for cell j. Thus, in this model we assume that the compensation to victims by the government is paid through the mandatory insurance.

The stability of the insurance program depends on whether the accumulated mutual fund together with the governmental compensation is able to cover claims, i.e., on the probability of event:

$$e_1 = \tau \sum_j \pi_j + v \sum_j L_j^\tau - \sum_j \phi_j L_j^\tau \geq 0 \qquad (15.2)$$

The stability also depends on the willingness of individuals to accept premiums, i.e., with the probability of overpayments:

$$e_2 = \tau\pi_j - \phi_j L_j^\tau \geq 0, \ j = 1, \ldots, m \tag{15.3}$$

Apart from the compensation $v \sum L_j^\tau(x)$ the government arranges a contingent credit y with a fee q to improve[j] the stability of the mandatory insurance by transforming event (15.2) into (15.4):

$$e_3 = \tau \sum_j \pi_j + v \sum_j L_j^\tau - \sum_j \phi_j L_j^\tau + y - \tau q y \geq 0 \tag{15.4}$$

Here we assume that the mandatory insurance pays the fee $\tau q y$ and receives the credit y, whereas the government pays back the credit with the interest rate γ $y, \gamma > 1$.
The difference between compensation $v \sum\limits_j L_j^\tau$ and contingent credit y is significant: the outflow of fees is smooth, whereas the compensation of claims has a sudden impact at time τ, and without y it may require a higher government compensation (greater v) possibly exceeding the available budget. Therefore, without ex-ante contingent injections of capital y the diversion of capital from other governmental needs may occur.

Inequalities (15.3) and (15.4) define important events, which constrain the choice of the decision variables specifying the insurance program, i.e., the compensation rate v by the government, coverage by the insurance company ϕ_j, premiums π_j, and credit y with fee q. The likelihood of events (15.3) and(15.4) and values e_2, e_3 determine the stability (resilience) of the program. In a rough way this can be expressed in terms of the probabilistic constraint

$$P[e_2 > 0, e_3 < 0] < p \tag{15.5}$$

where p is a desirable probability of the program's default, say a default that occurs only once in 100 years. Constraint (15.5) is similar to the so-called insolvency constraint (Stone 1973), a standard for regulations of the insurance business. In the stochastic optimization (Ermoliev and Wets 1988) constraint (15.5) is known as the so-called chance constraint. The main goal can now be formulated as the minimization of expected total losses $F(x) = E \sum (1 - \phi_j) L_j^\tau + \gamma y$ including uncovered (uninsured) losses by the insurance contracts and the cost of credit γy, subject to the chance constraint (15.5), where vector x consists of the components π_j, ϕ_j, y.

Constraint (15.5) imposes significant methodological challenges even in cases when $\tau(x)$ does not depend on x and events (15.3) and (15.4) are defined by linear functions of decision variables (see discussion in Ermoliev and Wets 1988, p. 8, and in Ermoliev et al. 2000a, b, 2001). This constraint is of "black-and-white" character, i.e., it accounts only for a violation of (15.3) and (15.4) and not for its size. There are important connections between the minimization of $F(x)$ subject to highly - non-linear and possibly discontinuous chance constraints (15.5) and the minimization

of convex functions, which have important economic interpretation. Consider the
following function

$$G(x) = F(x) + \alpha E \max\left\{0, \sum_j \phi_j L_j^\tau - v \sum_j L_j^\tau - \tau \sum_j \pi_j - y + \tau q y\right\}$$
$$+ \beta E \sum_j \max\left\{0, \tau \pi_j - \phi_j L_j^\tau\right\}, \tag{15.6}$$

where α, β are positive parameters.

It is possible to show (see chapter 2 in Ermoliev and Wets 1988 and more general
results in Ermoliev et al. 2000a, b, 2001) that for large enough α, β a minimization
of function $G(x)$ generates solutions x with $F(x)$ approaching the minimum of $F(x)$
subject to (15.5) for any given level p.

The minimization of $G(x)$ defined by (15.6) has a simple economic interpretation. Function $F(x)$ comprises expected direct losses associated with the insurance
program. The second term includes the expected shortfall of the program to fulfill
the obligations; it can be viewed as the expected amount of ex-post borrowing with
a fee α. Similarly, the third term can be interpreted as the expected ex-post
borrowing with a fee β needed to compensate overpayments. Obviously that large
enough fees α, β will tend to preclude the violation of (15.3) and (15.4). Thus, the ex-
post borrowing with large enough fees allows for a control of the insolvency
constraints (15.5). It is easy to see that the use of the ex-post borrowing (expected
shortfall) in the second term of $G(x)$ in combination with the optimal ex-ante
contingent credit y controls the CVaR type risk measures. Indeed, the minimization
of $G(x)$ is an example of stochastic minimax problems (see Ermoliev and Wets
1988, chapter 22). By using standard optimality conditions for these problems we
can derive the optimality conditions for the contingent credit y. For example,
assuming continuous differentiability of $G(x)$ which follows in particular from
the continuity of underlying probability distributions, it is easy to see that the
optimal level of the credit $y>0$ satisfies the equation

$$\frac{\partial G}{\partial y} = \gamma - \alpha P\left[\sum_j \phi_j L_j^\tau - v \sum_j L_j^\tau - \tau \sum_j \pi_j > y\right] = 0 \tag{15.7}$$

Thus, the optimal amount of the contingent credit is defined as a quantile of the
random variable $\sum_j \phi_j L_j^\tau - v \sum_j L_j^\tau - \tau \sum_j \pi_j$ specified by the ratio γ/α, which has to
be not greater than 1. Hence, the expectation in the second term of $G(x)$ for optimal y
is taken under the condition that y is the quantile of $\sum_j \phi_j L_j^\tau - v \sum_j L_j^\tau - \tau \sum_j \pi_j$. This
is in accordance with the definition of CVaR (Artzner et al. 1999; Rockafellar and
Uryasev 2000). More general risk measures emerge from the optimality conditions
of $G(x)$ with respect to premiums π_j, ϕ_j.

The importance of using expected shortfall as an economically sound risk
measure was emphasized by many authors (see Arrow 1996; Embrechts et al.

2000; Jobst and Zenios 2001; Rockafellar and Uryasev 2000). Important connections of CVaR with the linear programs were discussed e.g. in Rockafellar and Uryasev 2000. Let us note that $G(x)$ is a convex function in the case when τ and L_j^τ do not depend on x. In this case the stochastic minimax problem (15.6) can be approximately solved by linear programming methods (see general discussion in Ermolieva et al. 1997). The main challenge is concerned with the case when τ and L^τ are implicit functions of x. Then we can only use the Adaptive Monte Carlo optimization. Let us outline only the main idea of these techniques. More details and further references can be found in Ermoliev et al. 2000a, b.

Let us assume that vector x incorporates not only risk management decision variables but also includes components affecting the efficiency of the sampling itself (for more detail see Ermolieva 1997; Ermoliev et al. 2001). An adaptive Monte Carlo procedure searching for a solution minimizing $G(x)$ of type (15.6) starts at any reasonable guess x^0. It updates the solution sequentially at steps $k = 0$, $1, \ldots$, by the rule $x^{k+1} = x^k - \rho_k \xi^k$, where numbers $\rho_k > 0$ are predetermined step-sizes satisfying the condition $\sum_{k=0}^{\infty} \rho_k = \infty$, $\sum_{k=0}^{\infty} \rho_k^2 = \infty$. For example, the specification $\rho_k = 1/k + 1$ would suit. Random vector ξ^k is an estimate of the gradient $G_x(x)$ or its analogs for non-smooth function $G(x)$. This vector is easily computed from random observations of $G(x)$. For example, let G^k be a random observation of $G(x)$ at $x = x^k$ and $\tilde{G}^k$ be a random observation of $G(x)$ at $x = x^k + \delta_k h^k$. The numbers δ_k are positive, $\delta_k \to 0$, $k \to \infty$, and h^k is an independent observation of the vector h with independent and uniformly distributed on $[-1, 1]$ components. Then ξ^k can be chosen as $\xi^k = \left[\left(\tilde{G}^k - G^k \right) / \delta_k \right] h^k$. The formal analysis of this method, in particular, for discontinuous goal functions, is based on general ideas of the stochastic quasi-gradient methods (see Ermoliev and Wets 1988 and further references in Ermoliev et al. 2001).

15.5 Demand for Ex-Ante Contingent Credit

According to the model of Sect. 15.4 and optimality condition (15.7), the demand of the pilot region for a contingent credit significantly depends on the composition of other risk management options and on the interplay of various factors. First of all it depends on the occurrence of floods, the reliability of the flood protection system and the prevailing land-use practices affecting the discharge curves. The demand for a contingent credit increases with difficulties in raising ex-post credits and governmental compensations.

In the following we discuss some numerical experiments using real data collected in the Upper Tisza region. For the illustration of the proposed model we use some simplified assumptions. The case study region is subdivided into $1,500 \times 1,500$ grids of 10 m^2, and for each grid there are data on the vulnerability of its assets. These grids are further aggregated into 40 cells. In the numerical experiments we analyzed the demand for the credit under two alternative

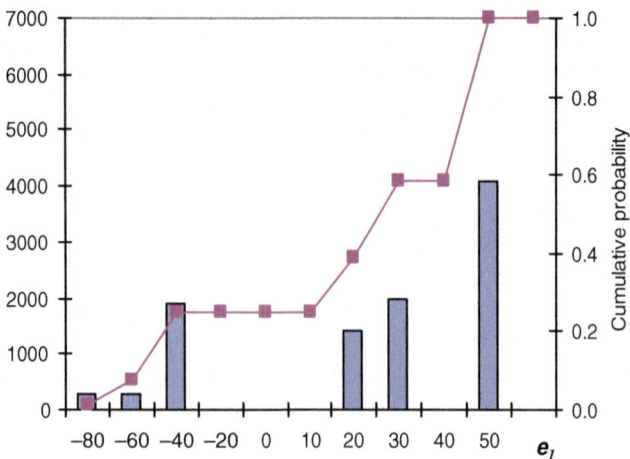

Fig. 15.3 Actuarial premiums

assumptions on premiums π_j, i.e., actuarial premiums calculated based on available data on average losses without differentiating the spatial risks and "fair" premiums derived out of the equilibrium conditions (15.3), (15.4), where $j = 1, \ldots, m$ and m is a number of cells/locations, in our case $m = 40$. The flood occurrences are modeled according to specified probabilities of rainfall patterns and dike breaks. There are three dikes allocated along the pilot river branch. Each of them may break under the probability of a 100, 150, and 1,000-year flood. In this chapter we take into account only structural losses. The simulation time horizon is assumed to be 50 years. The number of simulations (scenarios) in a single experiment was 10,000. A contingent credit in our model is introduced to stabilize Eq. (15.2) according to Eq. (15.4). The demand for the credit is, therefore, defined by negative values of indicator e_1 or e_3 for optimal solutions ϕ_j, π_j minimizing (15.6) for $y = 0$ and given v. This defines also the lack of capacity for the mandatory insurance. Figures 15.3 and 15.4 illustrate the results of the experiments with $v = 0.25$, $y = 0$. The horizontal axis shows the total demand for contingent credit, negative e_1, whereas the vertical axis shows the number of simulations and the cumulative probability.

In practical calculations (see Amendola et al. 2000a, b; Ermoliev et al. 2000a, b and Sect. 15.5) histograms for constraints (15.5) calculated simultaneously with the minimization of (15.6) provide a signal for increasing or decreasing risk factors α, β to achieve a satisfactory level p. Intuitively, greater α, β lead to constraints (15.5) with smaller p. On the other hand, this may considerably reduce insurance coverage of catastrophic exposures. A trade-off between these two effects can be resolved by using some additional considerations, e.g., political considerations or purely visual character of histograms, which cannot be formalized in general within a single model. Analysis of outcomes generated by alternative risk factors is similar to the standard welfare analysis.

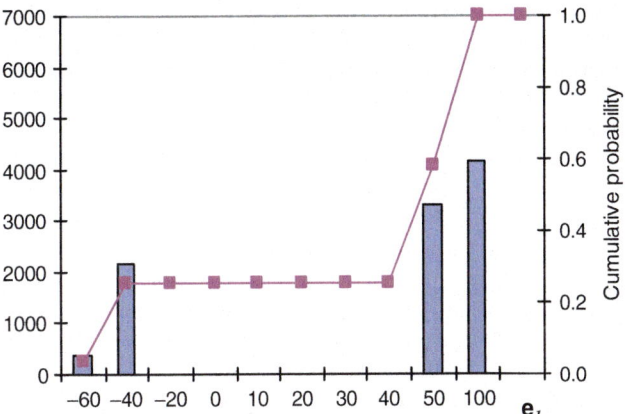

Fig. 15.4 Optimal premiums

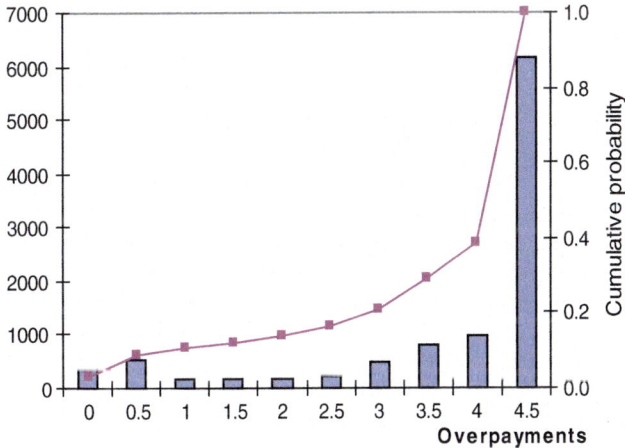

Fig. 15.5 Actuarial premiums

According to our experiments, the premium for the first option equals on average (per location and year) 0.87 million HF (Hungarian forints) (exchange rate: one HF equals 0.003302 US dollars). As we can see from Fig. 15.3, the inflow of premiums is not enough to compensate the losses, since e_1 (which defines a certain safety (solvency) level p for constraint in (15.5)) is often negative. In more than 2,000 scenarios out of 10,000 simulated catastrophic events the mandatory insurance at the given premium lacks the capacity to cover losses. This calls for a more significant intervention by the government through either increasing the level of compensation v, and/or through contingent credit.

Location-specific optimal premiums improve the situation. Figure 15.4 illustrates the changes in the total demand for contingent credit by using the optimal

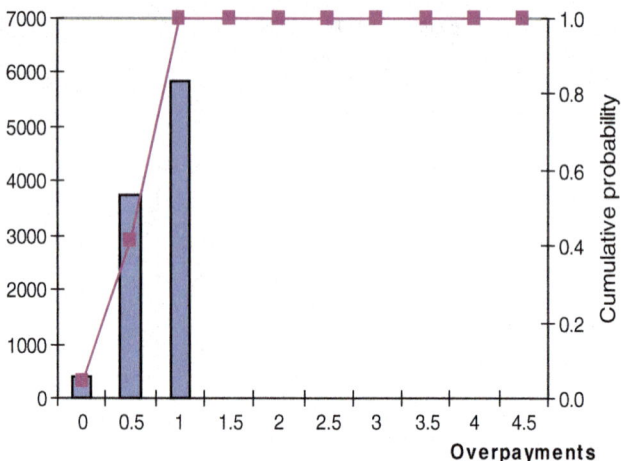

Fig. 15.6 Optimal premiums

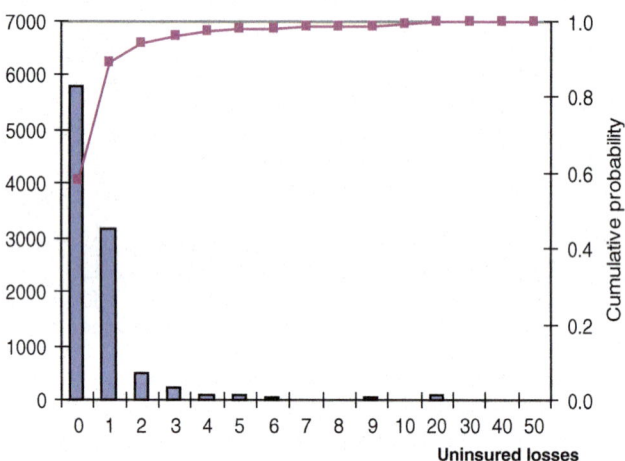

Fig. 15.7 Actuarial premiums

premiums calculated from the minimization of (15.6) for the same $v = 0.25, y = 0, \alpha$ and β. The model suggests a premium rate on average (per location and year) equal to 0.83 million HF, which is lower than in the first case. Figure 15.4 shows that the demand for contingent credit is also reduced (fewer negative values on the horizontal axis).

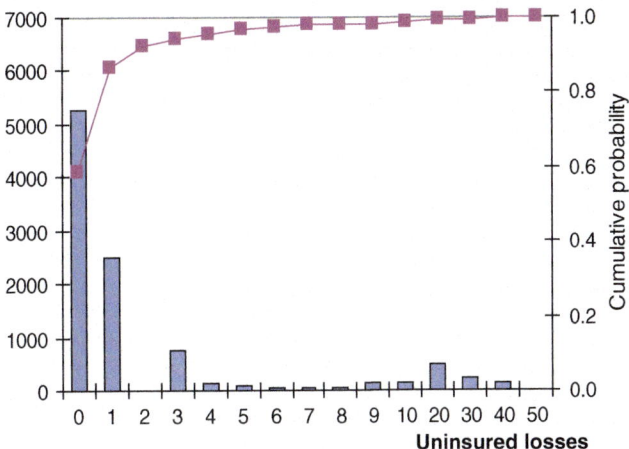

Fig. 15.8 Optimal premiums

Optimal premiums improve also overpayments per year (Figs. 15.5 and 15.6), i.e., the distribution of $\sum_j \max\left\{0, \tau\pi_j - \phi_j L_j^\tau\right\}/\tau$ computed from the third term of $G(x)$. Figures 15.7 and 15.8 show the distribution of uninsured losses computed from the first term of $G(x)$. It is evident (Fig. 15.8) that the optimal premiums reduce coverage in order to stabilize the insolvency (negative e_1 in Fig. 15.3) and overpayments. In these experiments we used $\alpha = \beta = N = 10$. Figures 15.8 and 15.9 show further reductions of overpayments and the demand for contingent credit for $N = 20$, which are, in fact, due to further reductions of coverage and premiums (Fig. 15.10).

The computer program (optimization part) was implemented in Matlab. The optimization procedure is easily restarted from different initial solutions, for new compositions of cells, and distributions of random parameters. The solution time slightly changes with the number of decision variables and random parameters. It may increase with the increase of N (unreasonably large N may cause degeneracy of G – level sets) and it also depends on the frequency of catastrophes. An important idea to reduce this time is to use faster versions of the Monte Carlo simulations.

15.6   Concluding Remarks

In this chapter we argue that because of significant interdependencies among catastrophe losses across different locations, the demand for a particular financial instrument cannot be separated from the demand for other risk transfer and risk reduction measures. We demonstrated that the demand for a specific contingent risk transfer instrument, a contingent credit, significantly depends on other pillars of the loss-spreading program for the pilot region of the Upper Tisza river, Hungary. In

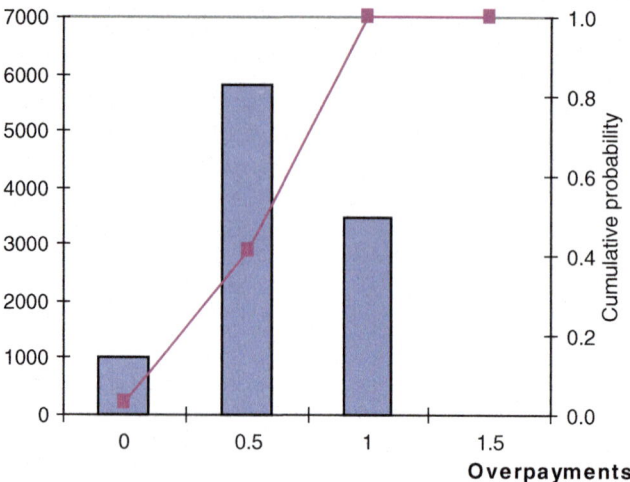

Fig. 15.9 Risk coefficient $N = 20$

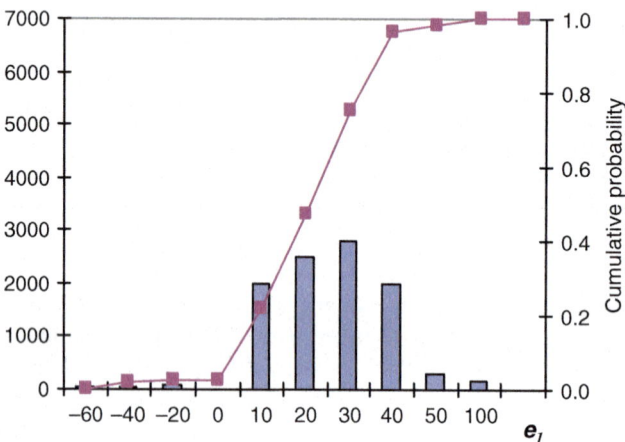

Fig. 15.10 Risk coefficient $N = 20$

particular, our numerical experiments show that optimal location-specific premiums of the mandatory insurance decrease the demand for contingent credit. This analysis may be extended by using real data from the region to derive the optimal size of contingent credit under different risk reduction measures, e.g., by strengthening or removing some of the dikes. Preliminarily, purely toy exercises show that the diversion of capital from direct compensation to investments into loss reduction measures may essentially reduce dependencies among losses and, hence, stabilize the mandatory insurance and reduce further the demand for cross-subsidies on country-wide and international levels. It is also important to analyze the effects of adaptive sampling procedures (adaptive Monte Carlo simulations), i.e., to enrich the set of decisions by components controlling the efficiency of the sampling. Traditionally, the insurance industry manages independent risks by pooling its exposures through contracts written on the basis of rich historical data. This is not possible with regard to rare catastrophic risk with mutually dependent losses of high consequences. Our model incorporates a generator of catastrophic losses (catastrophe model) and it is able to optimize effects of dependent and location-specific rare risks through iterative revisions of the catastrophe model by a stochastic optimization procedure. An essential challenge is to develop a special version of the catastrophe model enabling fast simulation of catastrophes and comparative analysis of management decisions.

The reduction of highly nonlinear and often discontinuous insolvency constraint (15.5) to a minimization of a non-differentiable and, in our case, a convex function (15.6) is a rather promising idea. The minimization of function (15.6) can be viewed as the simplest version of the so-called two-stage stochastic programs (see chapter 2 of Ermoliev and Wets 1988) and stochastic minimax problems, which are usually solved by linear programming methods. The dependency of (15.6) on the stopping time generally destroys concavity and even continuity of goal functions and brings new challenges only briefly outlined in this chapter.

References

Amendola A, Ermoliev Y, Ermolieva T, Gitits V, Koff G, Linnerooth-Bayer J (2000a) A systems approach to modeling catastrophic risk and insurability. Nat Hazards 21(2–3):381–393

Amendola A, Ermoliev Y, Ermolieva T (2000) Earthquake risk management: a case study for an Italian region. In: Proceedings of the second Euroconference on global change and catastrophe risk management: earthquake risks in Europe. IIASA, Laxenburg, Austria, 6–9 July 2000

Arrow K (1996) The theory of risk-bearing. small and great risks. J Risk Uncertain 12:103–111

Artzner P, Delbaen F, Eber JM, Heath D (1999) Coherent measures of risk. Math Financ 9(3):203–228

Cummins J D, Doherty N (1996) Can insurers pay for the "big one"? Measuring capacity of an insurance market to respond to catastrophic losses. Working paper, Wharton risk management and decision processes center, University of Pennsylvania, Philadelphia

Dilley M, Chen R, Deichmann U, Lerner-Lam A, Arnold M with Agwe J, Buys P, Kjekstad O, Lyon B, Yetman G (2005) Natural disaster hotspots: a global risk analysis. Disaster risk management series 5. The World Bank Hazard Management Unit, Washington, DC

Embrechts P, Klueppelberg C, Mikosch T (2000) Modeling extremal events for insurance and finance. Applications of mathematics, stochastic modeling and applied probability, vol 33. Springer, Heidelberg

Ermoliev Y, Wets R (1988) Numerical techniques of stochastic optimization. Computational mathematics. Springer, Berlin

Ermoliev Y, Ermolieva T, MacDonald G, Norkin V (2000a) Insurability of catastrophic risks: the stochastic optimization model. Optim J 47:251–265

Ermoliev Y, Ermolieva T, MacDonald G, Norkin V (2000b) Stochastic optimization of insurance portfolios for managing exposure yo catastrophic risks. Ann Oper Res 99:207–225

Ermoliev Y, Ermolieva T, MacDonald G, Norkin V (2001) Problems on insurance of catastrophic risks. Cybern Syst Anal 2:90–110

Ermolieva T (1997) The design of optimal insurance decisions in the presence of catastrophic risks. International Institute for Applied Systems Analysis Interim Report IR-97-068, Laxenburg, Austria

Ermolieva T, Ermoliev Y, Norkin V (1997) Spatial stochastic model for optimization capacity of insurance networks under dependent catastrophic risks: numerical experiments. International Institute for Applied Systems Analysis Interim Report IR-97-028, Laxenburg, Austria

Froot K (1997) The limited financing of catastrophe risk: an overview. Harvard Business School/National Bureau of Economic Research, Cambridge

Jobst N, Zenios S (2001) The tail that wags the dog: integrating credit risk in asset portfolios. J Risk Financ 3(1):31–43

Kleindorfer P, Kunreuther H (1999) The complementary roles of mitigation and insurance in managing catastrophic risks. J Risk Anal 19:727–738

Kunreuther H, Linnerooth-Bayer J (2000) The financial management of catastrophic flood risks in emerging economy countries. In: Proceedings of the second Euroconference on global change and catastrophe risk management: earthquake risks in Europe. IIASA, Laxenburg, Austria, 6–9 July 2000

Mayers D, Smith C (1983) The interdependencies of individual portfolio decisions and the demand for insurance. J Pol Econ 91(2):304–311

Pollner J (2000) Catastrophe risk management: using alternative risk financing and insurance pooling mechanisms. Finance, private sector & infrastructure sector unit, Caribbean Country Department, Latin America and the Caribbean region, World Bank

Project Proposal (2000) Project on flood risk management policy in the upper Tisza basin: a system analytical approach. Risk, modelling and society project. International Institute for Applied Systems Analysis, Laxenburg, Austria

Pugh EL (1966) A gradient technique of adaptive Monte Carlo. SIAM Rev 8(3):346–355

Rockafellar T, Uryasev S (2000) Optimization of conditional value-at-risk. J Risk 2(3):21–41

Stone JM (1973) A theory of capacity and the insurance of catastrophe risks. Parts 1, 2. J Risk Insur 40:231–244, 339–355

Walker G (1997) Current developments in catastrophe modelling. In: Britton NR, Oliver J (eds) Financial risks management for natural catastrophes. Griffith University, Brisbane, pp 17–35

Chapter 16
Flood Risk in a Changing Climate: A Multilevel Approach for Risk Management

Stefan Hochrainer-Stigler, Georg Pflug, and Nicola Lugeri

Abstract Many regions in Europe are vulnerable to the expected increase in the frequency and intensity of climate related hazards, and partly for this reason adaptation to climate change is moving to the forefront of EU and national policy. Yet, little is known about the changing risks of floods and other weather hazards, or about possible adaptation options under these dynamic conditions. Based on a risk modeling approach this chapter presents results on estimating changing flood losses for Hungary, including the Tisza region, which is highly exposed to flooding. We examine two generic options for managing the risks, including investments in risk reduction measures and investments in insurance or other risk-transfer instruments. The analysis distinguishes between different layers of risk (high and low consequence), which provides insights on allocating between risk reduction and risk transfer.

Keywords Flood risk • Climate change • Adaptation • Risk layer approach • Hungary • Tisza region

16.1 Introduction

Europe is vulnerable to current and future weather related disaster risk, and climate adaptation is moving to the forefront of EU and national policy (Kundzewicz and Mechler 2010). Yet, little is known about changing risks under dynamic conditions and the corresponding advantages and limits of options for adapting to climate

S. Hochrainer-Stigler (✉) • G. Pflug
International Institute for Applied Systems Analysis (IIASA),
Schlossplatz 1, A-2361 Laxenburg, Austria
e-mail: hochrain@iiasa.ac.at

N. Lugeri
Istituto Superiore Protezione e Ricerca Ambientale, Rome, Italy

A. Amendola et al. (eds.), *Integrated Catastrophe Risk Modeling: Supporting Policy Processes*, Advances in Natural and Technological Hazards Research 32,
DOI 10.1007/978-94-007-2226-2_16, © Springer Science+Business Media Dordrecht 2013

change. This chapter focuses on Hungary and especially the Tisza region, which are severely exposed to current and future flood risk (Linnerooth-Bayer et al. 2001; Kovacs 2006; Parry et al. 2007).

Based on the EU research project ADAM (Adaptation and Mitigation Strategies, see Kundzewicz and Mechler 2010), this chapter assesses how climate change can affect flood risks in Hungary and the Tisza region, and discusses potential policy options for its management. The assessment is based on the CATSIM approach (see Chap. 8 in this book), and consistent with this approach the results are presented in terms of (monetary) loss distributions. Current and (partly) future losses (projected to 2071–2100) are estimated taking account of the hazard, the exposure of assets as well as their physical vulnerability. The risk to human life is not included.

Unique to this analysis, changes in anticipated losses, compared to the current risk situation, are separated into different risk layers. A low risk layer is comprised of relatively frequent and low-consequence events, e.g., the 20- to 100-year events, while a high risk layer includes loss events with low probability but high consequences, e.g., the 100- to 500-year events. These different risk layers can be expressed simultaneously with loss distributions or loss exceedance curves. Since risk-reduction and risk-transfer measures are suitable for different risk layers, this distinction should help the decision maker to determine the most appropriate adaptation measures. This approach is especially useful if dissimilar changes in risk at the local level can be expected, e.g., an increase in low-probability events in one region and an increase in the frequency of small magnitude events in a neighbouring region.

This chapter is organized as follows: In Sect. 16.2 current flood risk estimates at the country and regional levels are presented, and in Sect. 16.3 the risks are estimated for the year 2100 taking account of climate change. An adaptation methodology to lessen losses taking account of risk layers is presented in Sect. 16.4. Finally, Sect. 16.5 discusses the limitations of the method and ends with a conclusion and outlook for the future.

16.2 Direct Flood Risk at the National and Regional Level in Hungary

Direct flood losses are damages incurred from the immediate impact of the flood, for example, to physical infrastructure and buildings. The term *direct flood risk* characterizes the relationship between direct flood losses and the probability of such events, usually in the form of a loss distribution (LD) curve or a loss exceedance probability (EP) curve. An annual LD curve indicates the probability p that *at most* \$X is lost in a given year, while an EP curve indicates the probability p that *at least* \$X is lost in a given year. Hence, the LD and EP curves provide the same information (Hochrainer 2006); however, in the catastrophe modeling litera-ture the latter is used (Grossi and Kunreuther 2005) while in the risk management

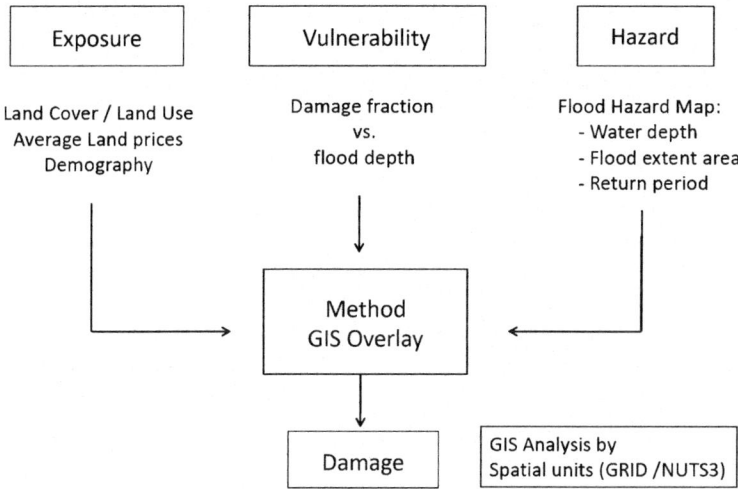

Fig. 16.1 Scheme for the evaluation of flood risk (Source: Adapted from Lugeri et al. 2009)

literature the (classical) LD curve is used (Pflug and Römisch 2007). Catastrophe models (Compton et al. 2009), including CATSIM, are discussed in Chap. 8 of this book and therefore not discussed in detail here. The following section describes the approach applied to estimate flood losses in Hungary and the Tisza region.

16.2.1   Direct Flood Risk Estimation Methodology

Catastrophe models (see also Chap. 8) combine three components: hazard, exposure, and physical vulnerability. The results are usually in the form of pixel-based damages and their corresponding probabilities.

Based on Lugeri et al. (2009, 2010), we use a hybrid up-scaling method to aggregate the loss curves from the pixel to regional level. Figure 16.1 shows our approach for estimating direct flood risk at the European scale.

To compute *exposure* the CORINE (Co-ordination of Information on the Environment) Land Cover (CLC) map of the European natural and artificial landscape is applied (EEA 2000). The combination of hazard and land cover maps allows the estimation of the exposure to floods, which are grouped according to the classes of the land cover map. Generally, damage caused by floods depends on the main characteristics of the flood, including its duration, water depth, and water flow velocity. However, given the limitations of available data across Europe, it was only possible to quantify these effects by considering the flood depth alone. Therefore, the vulnerability of the assets under threat is estimated by means of depth-damage functions for relevant land use classes of CLC for all EU 27 countries.

Flood *hazard* maps are generated from catchment characteristics. The main component is the Digital Terrain Model (DTM), which allows an evaluation of the water depth in each location after a given flood. The flood hazard map was obtained from a 1 km * 1 km grid DTM and the dataset of European flow networks with the same grid size. The latter was developed by the Natural Hazard Action Group at the Joint Research Center (JRC) (see Barredo et al. 2005). It provides quantitative information on both the expected extent of flooded area and water depth. Since the map is based solely on topographic features, an algorithm has been developed to compute the height difference between a specific grid-cell and its closest neighboring grid-cell containing a river, while respecting the catchment tree-structure. Hence, the hazard map provides "static" information, differently from hydrological model-based maps that provide probabilistic information arranged in terms of return periods or probability of exceedance. A probabilistic read-out of the topography-based hazard map was made possible making use of expert judgment and calibration from the LISFLOOD hydrologic model (de Roo 1998), which is available for some catchments of Europe (Lugeri et al. 2009). The difficulty (due to the level of details needed and data availability) of using hydrological models on such a macro scale constitutes a major limitation, which is accounted for by including, where possible, uncertainty ranges in terms of maximum and minimum loss values.

The computation of monetary flood risk relies on GIS processing of hazard maps and territorial databases (the CLC and the stage-damage functions) combining the probability and severity of flooding with spatially explicit information on exposure and vulnerability. Model outcomes are grid based maps (pixel resolution 250 m * 250 m) of monetary damage computed for five return periods (50, 100, 250, 500 and 1,000 years). The basic maps are further elaborated to produce average annual loss estimates (via integration of the EP curve).

The spatial aggregation of the results from the GRID scale to the regional and country level is made possible by the hybrid convolution approach that was expressly designed within the ADAM project (Hochrainer et al. 2012). This approach is based on the idea that, depending on the magnitude of the hazard, losses in different elementary spatial units (clusters) and their aggregation are assumed to be co-monotonic, i.e. up to a given magnitude of the hazard and a given size of the group of clusters, the losses are assumed to be totally independent (e.g., occurrence of one event in one cluster does not alter the probability that it also occurs in another cluster); afterwards they are assumed to be fully dependent (Fig. 16.2). As the combination of two independent distributions is called "convolution", and the combination of two totally dependent distributions is simply the sum of them, this technique is called "Hybrid Convolution".

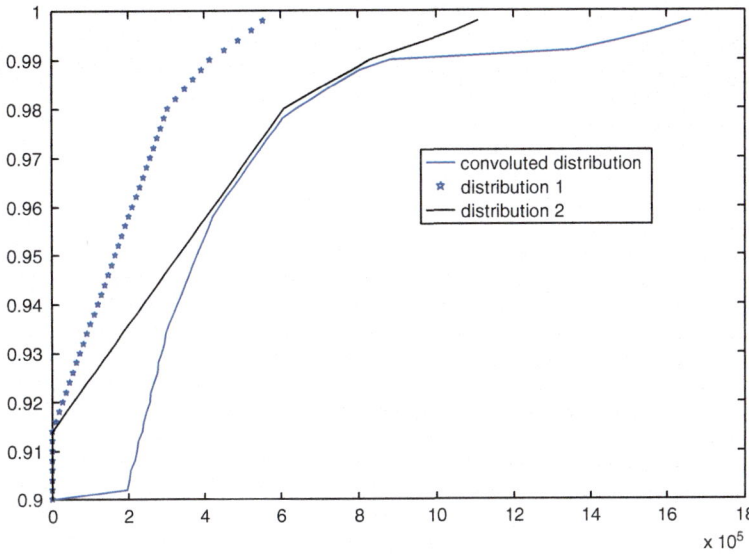

Fig. 16.2 Hybrid convolution of two distributions assuming total dependence after the 100-year event and no losses up to the 10-year event. Before the 100-year event (0.99 on the y axis) the two distributions are assumed to be totally independent and afterwards fully dependent

The clustering scheme as the basic spatial aggregation frame is defined according to the river catchment structure. In order to achieve a national breakdown of this catchment-based aggregation method, the calculations are performed nationwide by selecting the elementary clusters belonging to each administrative unit at the chosen NUTS[1] level (Fig. 16.3).

Using this method, eight cluster levels were defined, where the first cluster represents the smallest sub-catchment corresponding to a single drainage branch in the river network, and the eighth or last cluster is the largest level (country scale). As Fig. 16.3 indicates, the clusters are structured hierarchically with each cluster in each level being disjoint (Lugeri et al. 2010). In this manner proceeding from lower clusters to higher clusters via the hybrid convolution process, loss distributions with the desired resolution are obtained.

---

[1] NUTS (Nomenclature of Territorial Units for Statistics) is a geocode standard for referencing the subdivisions of countries. Usually, a hierarchy of three NUTS levels for each EU member country is established. Subdivisions usually (but not necessarily) correspond to administrative divisions within the country (see Lugeri et al. 2010).

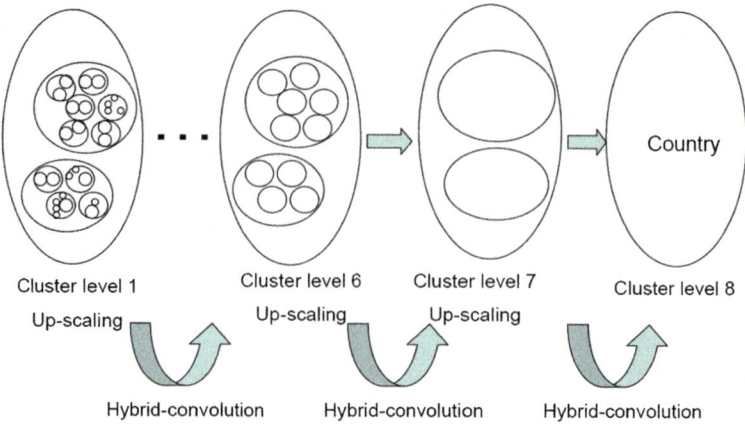

Fig. 16.3 Upscaling process from lower to higher spatial levels (defined here as cluster levels) (Source: Lugeri et al. 2010)

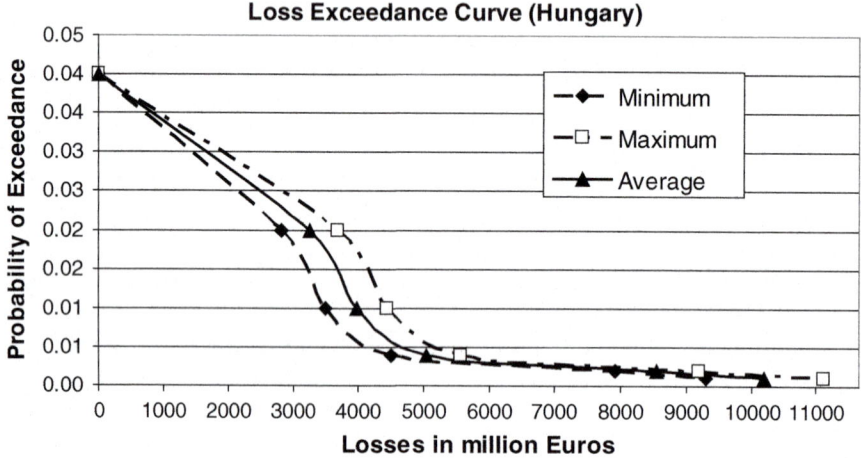

Fig. 16.4 Annual loss exceedance curve for Hungary (Source: Hochrainer and Mechler 2009)

16.2.2 Flood Risk in Hungary

Based on the approach described above the results are presented and extended to include the regional and national levels. Figure 16.4 shows the estimated current flood loss exceedance curve for Hungary.

The area under the loss exceedance curve yields the expected annual loss, which is estimated to be around 128 million €. However, due to modelling issues within the hazard component, there is considerable uncertainty around this estimate. Using

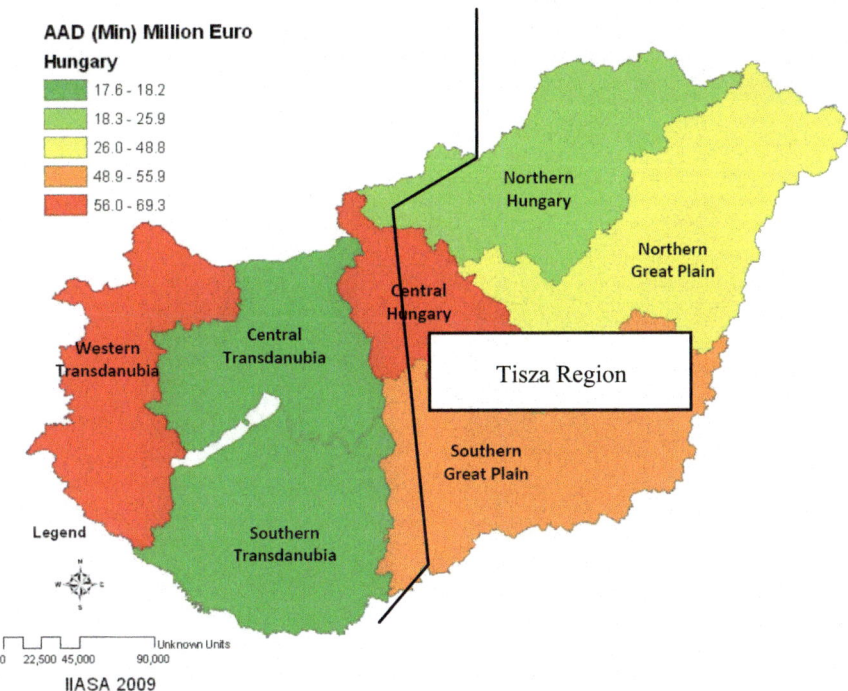

Fig. 16.5 Average annual losses for regions in Hungary (Source: Own calculations)

upper and lower bounds (based on the hazard analysis) the minimum and maximum average loss is estimated to be between 114 and 143 million €, respectively. At the regional level (or NUTS2 level), one can see that estimated losses are especially large in the Tisza regions (eastern Hungary), including parts of Central Hungary, Northern Hungary, the Northern Great Plain and Southern Great Plain (Fig. 16.5).

While estimates of average losses are one indicator for risk, they do not capture extreme events or the heavy tail of the loss exceedance curve for which other measures are necessary (Pflug and Römisch 2007). Figure 16.6 illustrates the 100-year loss for the respective regions in Hungary. As can be seen from this figure, some of the Tisza regions can expect losses over 1 billion € for the 100-year event.

16.3   Estimating Future Flood Losses in the Tisza Regions Taking Account of Climate Change

Climate change will affect potential climatic hazards in the future and therefore change the direct risk either by increasing/decreasing the frequency and/or intensity of events in the future (Parry et al. 2007; Dankers and Feyen 2008). The effect of

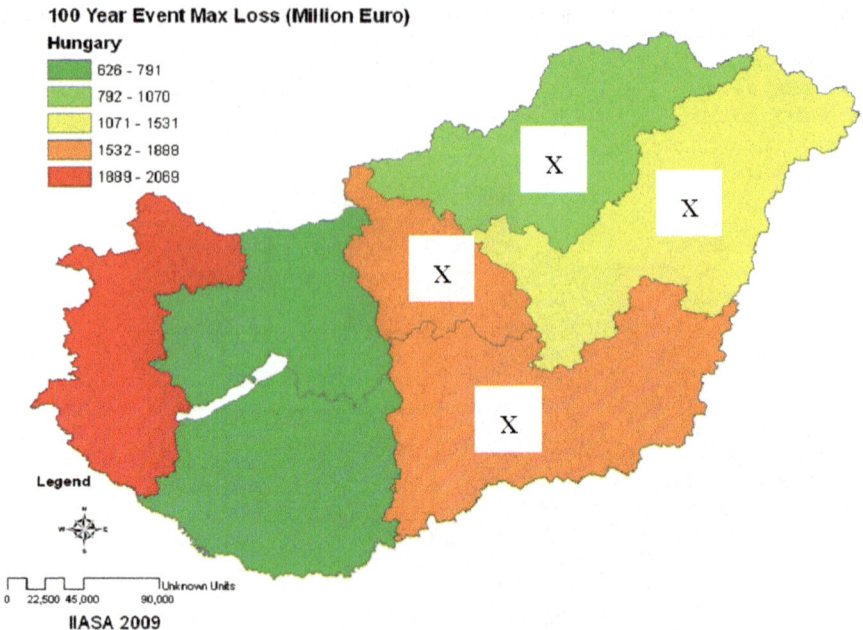

Fig. 16.6 100 year loss estimates for regions in Hungary (Source: Own calculations)

climate change on future disaster losses, however, is highly complex, and investigations are still in their infancy (Alcamo et al. 2007). We base our analysis on a recent study by Hirabayashi et al. (2008) to determine changes in flood hazard frequency in Hungary over the period 2010–2100 for a 100-year event for the A1B storyline. The A1B storyline is a sub-scenario within the A1 storyline and defined as follows:

> The A1 storyline and scenario family describes a future world of very rapid economic growth, global population that peaks in mid-century and declines thereafter, and the rapid introduction of new and more efficient technologies. Major underlying themes are convergence among regions, capacity building, and increased cultural and social interactions, with a substantial reduction in regional differences in per capita income. The A1 scenario family develops into three groups that describe alternative directions of technological change in the energy system. The three A1 groups are distinguished by their technological emphasis: fossil intensive (A1FI), non-fossil energy sources (A1T), or a balance across all sources (A1B) (IPCC 2000, p. 4).

Estimates of the relative changes in the 100-year event serve as the basis to estimate updated loss-return periods based on current risk data from Lugeri et al. (2009). In other words, the loss distributions are shifted according to the projected changes in the hazard frequency. The estimated change in the recurrence of the 100-year flood across Europe is illustrated in Fig. 16.7. We see from this analysis that in many regions the 100-year flood will occur as frequently as every 10 years at the end of the twenty first century.

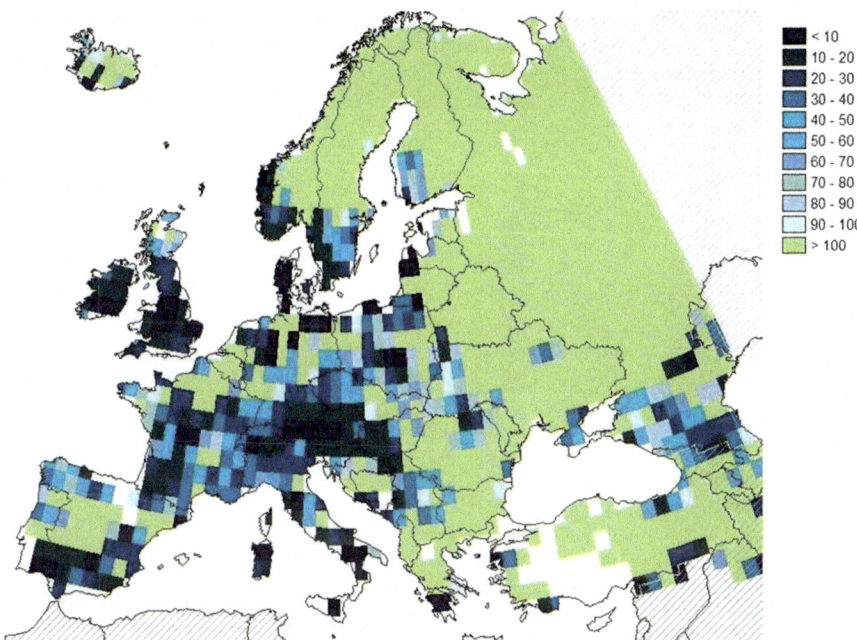

Fig. 16.7 Recurrence interval (return period) of today's 100-year flood (i.e. flood with a recurrence interval of 100 years during the period 1961–1990) at the end of the twenty first century (2071–2100), in case of scenario SRES A1B (Source: Kundzewicz et al. (2010), based on results from Hirabayashi et al. (2008))

It is not only climate change that will have an effect on future flood losses, but also other types of "global change" factors, such as economic development, land use and population dynamics, which will affect the elements at risk and the physical vulnerability of the assets. However, these factors cannot be included since changes in land use and population are scarcely predictable. Therefore, in this analysis we focus only on the flood hazard as it is likely to be affected by climate change using the already introduced hybrid convolution up-scaling technique to calculate (aggregated) losses at the regional level for 2071–2100. Estimates for the 50-, 100-, 250-, 500- and 1,000- year loss event where calculated, and Fig. 16.8 shows the results for the current and future risk situation for the 100-year loss event.

These results show a mixed picture for Hungary. The current 100-year loss for the Northern Great Plain was estimated to be 1,250 million €, and as shown in Fig. 16.8 this loss is expected to increase to 1,463 million € in 2071–2100. In contrast, for central Hungary the 100-year loss was estimated to be around 1,365 million €, which remains approximately constant to 2071–2100. For a better representation of the dynamics Table 16.1 illustrates the absolute changes for each region.

The Tisza region is represented here by Northern Hungary, the Northern Great Plain and Southern Great Plain, as well as part of Central Hungary. In Northern

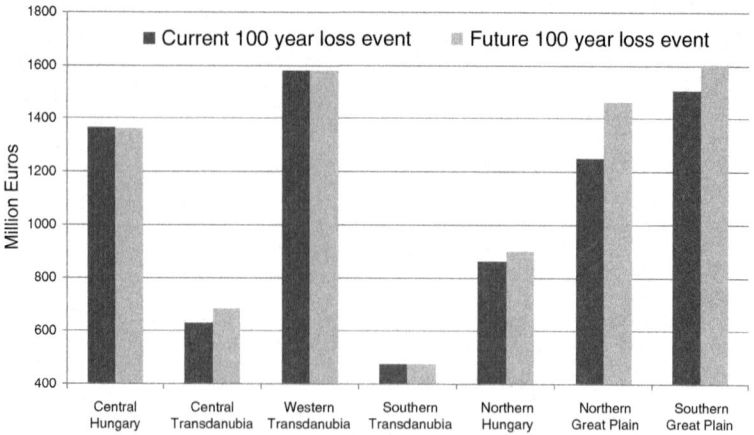

Fig. 16.8 Current and projected losses (2071–2100) for the 100-year event for the seven regional districts in Hungary

Table 16.1 Differences in losses (positive values showing an increase in losses) for different return periods with and without climate change for 2100 in million € for the seven regions in Hungary

Return period	50	100	250	500	1,000
Central Hungary	−20	−4	0	0	0
Central Transdanubia	42	54	149	317	3,554
Western Transdanubia	0	0	0	0	0
Southern Transdanubia	0	0	0	1	1
Northern Hungary	45	36	−132	−213	−115
Northern Great Plain	−4	215	880	2,176	4,143
Southern Great Plain	120	102	81	70	55

Hungary these estimates show a small increase in the losses for events with moderate frequency, but, interestingly, there is a decrease in losses for the very low probability (extreme) events. More drastic changes can be found for the Northern Great Plain with nearly all return periods showing a high increase in losses, especially for the very low probability events (the 500- and 1,000-year events). While all return periods show higher losses in the future in the Southern Great Plain, the increase of losses is less for low frequency events than for more frequent ones.

Summarizing, there are large regional differences in flood loss patterns in the future with an overall increase in absolute losses found in most regions. It should be kept in mind that without incorporating land use and other global change dimensions these results are only indicative. Still, without any decrease in exposure or vulnerability, the situation in the future is estimated to be worse than today in most of the cases. The next section takes a closer look at the changes from an adaptation point of view.

16.4   Risk Management Against Increases in Losses

Disaster risk management consists of measures taken before, during and after the disaster event to either reduce the extent of the hazard, exposure of the elements at risk and/or vulnerability (Lindell and Perry 2004). These measures can usefully be divided into risk-reduction (mitigation) instruments and risk-financing instruments. The former, such as levees to protect against floods, reduces direct hazard impacts, and the latter reduces the financial impacts. In addition, or alternatively, one can invest in insurance and other types of risk-financing measures to assure that sufficient resources are available after the event for timely relief and reconstruction. This reduces indirect "downstream" losses.

The choice between how much should be invested in risk reduction versus risk financing is not straightforward. Such a selection will depend primarily on the occurrence probability of the hazard, the associated size of impacts, the costs and benefits of both types of instruments, as well as on their interaction (see also Chap. 8). Risk-reduction measures against direct losses are usually more cost effective for frequent events (say 10–100-year loss return period) with low to medium-sized losses than for high impact/less frequent events. For lower impact events substantial reduction in damage can typically be achieved at relatively little expenditure, but it becomes increasingly costly to achieve further reductions in risk (Rescher 1983).

Concerning risk financing, risk bearers (individuals, businesses, governments) are generally better able to finance lower consequence events from their own means, including savings, reserve funds or credit arrangements. Risk financing is thus generally more appropriate for medium sized to extreme losses (say 100- to 500-year loss events) to smooth the *variability* of financial losses. Finally, some events (say above the 500-year loss event) can be too costly to be reduced or insured. Hence, outside assistance may be needed (e.g. from the government, international financial institutions such as the World Bank, donations, etc.).

If investments in risk reduction are more cost effective for low-consequence events, and investments in insurance more cost effective for high-consequence events, how would this translate into flood adaptation measures? Below (Fig. 16.9) we present an illustrative example of how a *risk layering* approach can provide insights on adaptation investments in the Tisza and other Hungarian regions.

We consider two stylized loss functions: the mitigation function and the insurance function. The consequence of these functions on the loss distribution can be expressed as follows: If a damage variable Y is distributed according to the distribution G, then the loss variable H(Y) is distributed according to the distribution function $G(H^{-1}(x))$. We employ the stylized mitigation function M as

$$M_{o,u}(x) = \begin{cases} 0 & \text{if } x \leq o \\ u(x-o)/(u-o) & \text{if } o \leq x \leq u \\ x & \text{if } x \geq u \end{cases}$$

Here "o" (the lower value) and "u" (the upper value) are parameters. The mitigation measure has the effect that hypothetical damage (i.e. without mitigation) smaller than the lower value "o" does not lead to financial loss. Hypothetical

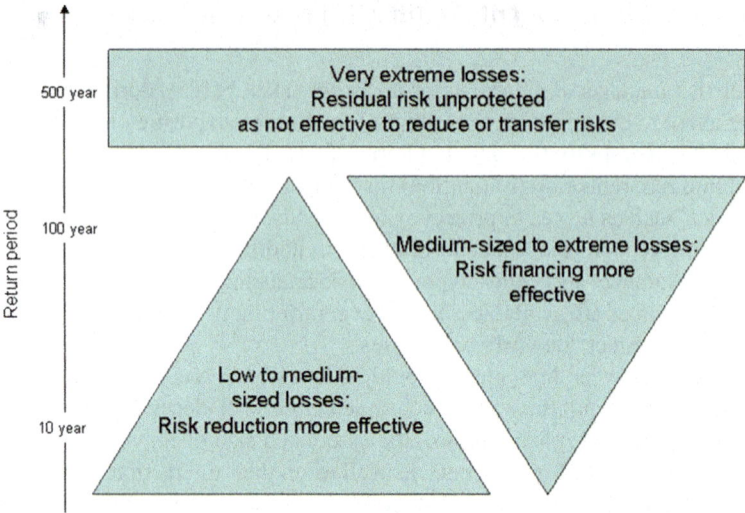

Fig. 16.9 Efficiency of risk management instruments and occurrence probability (Source: Mechler et al. 2009)

damages larger than the upper value "u" are too large to be mitigated and lead to full damage. In between, the function is linear.

As discussed, we assume that mitigation is more appropriate for smaller damages, and, in contrast, insurance more appropriate/feasible for large damages. We use here the function of a proportional excess of loss (XL-) insurance arrangement (often used for extremes, see also Chap. 8) with (monetary) attachment point "a", exit point "e" and proportionality factor "p", i.e. the insurer pays a proportion of losses starting from the attachment point till to the exit point. The stylized insurance function I is given by

$$I_{a,e,p}(x) = \begin{cases} 0 & \text{if } x \leq a \\ a + (x - a)(e - a)/b & a < x < e \\ x - (e - (a + b)) & \text{if } x \geq e \end{cases}$$
$$\text{with } b = (e - a)(1 - p)$$

As an illustrative example, in the top of Fig. 16.10 the dotted line is a mitigation function (with $o = 3{,}000$ and $u = 6{,}000$) whereas at the bottom of Fig. 16.10 the dotted line is an insurance function (with $a = 6{,}000$, $e = 10{,}000$ and $p = 0.5$).[2]

---

[2] The numbers chosen are only indicative and do not represent any "real" mitigation and insurance measure as this would need a comprehensive analysis on the very local level over all the regions which was not possible to perform due to data constraints. Hence, while the results may change using different numbers the approach would be still valid.

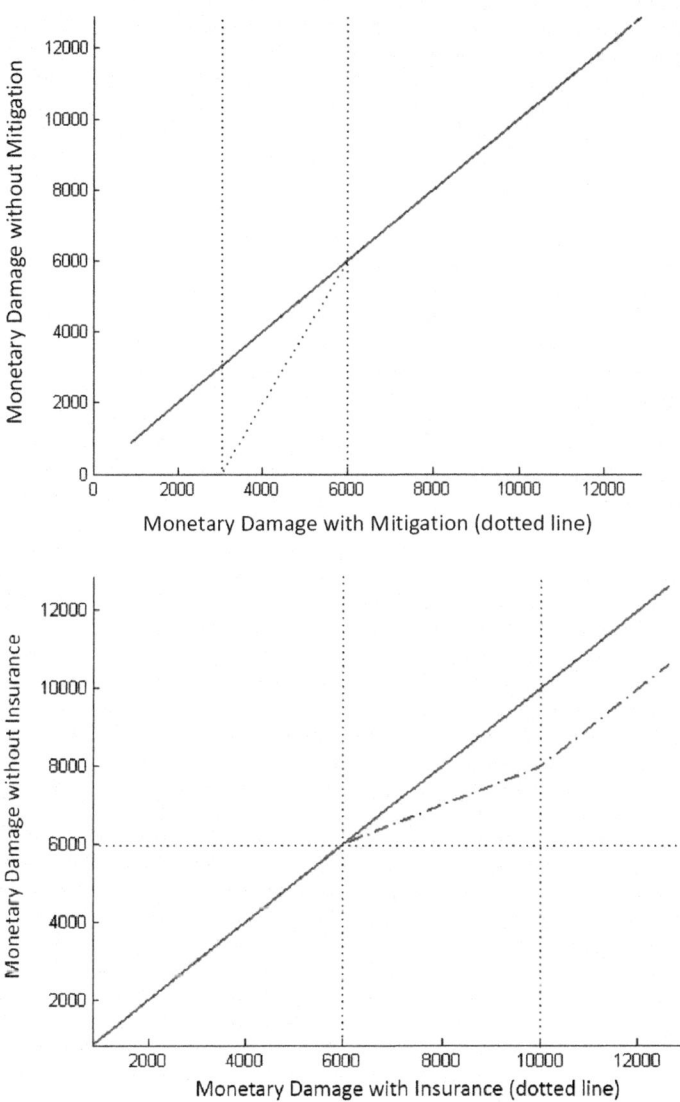

Fig. 16.10 Example of mitigation function (*top*) and insurance function (*bottom*)

Let F be today's damage distribution and G be the future damage distribution. The question we address is: which cost effective measures (mitigation and/or insurance) are needed so that the future distribution G is at most as risky as distribution F? It is important to be specific about what "riskiness" means, and we apply a conservative criterion, namely first order stochastic dominance (FSS), i.e. we want to have $G(x) \geq F(x)$ for all x, so that all decision makers (with

Table 16.2 Loading factors and corresponding p-quantiles. Based on Hochrainer 2006

Layer (Quantile)	Loading Factor
[0.000 0.850)	1.13
[0.850 0.947)	1.57
[0.947 0.965)	1.89
[0.965 0.975)	2.32
[0.975 0.985)	3.27
[0.985 0.992)	4.88
[0.992 0.993)	5.43
[0.993 0.996)	8.75
[0.996 1.000)	10.0

decreasing utility functions, i.e., the less the losses the better) would see the (adapted) future distribution G as less risky compared to F. In other words, our question can be stated as: what measures have to be used to decrease the numbers in Table 16.1 to zero (or smaller) with minimum costs? Hence, the optimization problem can be stated as

$$cost(o,u,G) + cost(a,e,p,M_{o,u} \circ G) \rightarrow Min$$
$$I_{a,e,p} \circ M_{o,u} \circ G \underset{FSS}{\geq} F$$
$$o, u, a, e, p \geq 0$$

As the optimization problem indicates, we first use mitigation to the necessary extent, afterwards we use this new distribution ($M_{o,u} \circ G$) for determining the necessary insurance parameters to achieve FSS.

Costs for each measure are determined from Mechler (2004) and MMC (2005), where cost-benefit ratios for flood defense measures were found to range from 1.5 till 4. Conservatively, we set the ratio to be 2, i.e., 1 € spent on mitigation yields 2 € in terms of mean reduction of losses in the selected range. To calculate the risk-adjusted insurance premium for a given insurance contract $I_{a,e,p}$ we use loading factors (called 'h') for different quantiles of the distribution F (see Hochrainer 2006) and calculate the premiums as

$$\pi(I_{a,e,p}, F) = \int_0^\infty I(y)h(F(y))dF(y)$$

The loading factors for each quantile interval are given in Table 16.2.

As an example of the results, we focus now on the Northern Great Plain which shows large increases in losses due to flood events in the future (see Table 16.1). As Fig. 16.11 indicates these increases correspond to a shift of the loss distribution to the right.

As there are increases not only in extreme events but also in more frequent ones, a combination of mitigation and insurance is necessary to shift the future distribution back to the current distribution (due to the mitigation model used, there is a decrease in risk at lower layers). To achieve this most cost efficiently, mitigation

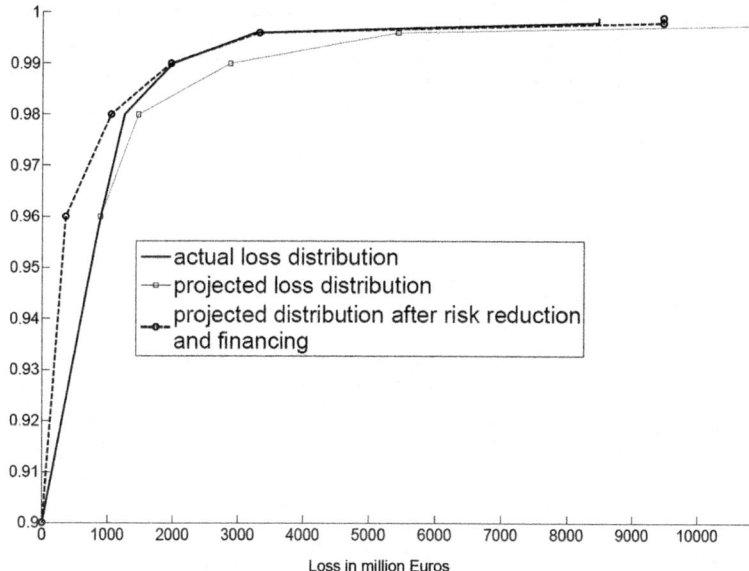

Fig. 16.11 Northern Great Plain: loss distributions for today and in the future (with and without adaptation)

Table 16.3 Mitigation and insurance parameters estimated for selected regions in Hungary

	Parameters					*Costs(million €)*	
	o	u	a	e	Prop	Mitig.	Ins.
Northern Great Plain	600	3,000	1,000	5,500	0.7	13.85	170
Central Transdanubia	200	3,000	1,000	5,500	0.9	5.86	76
Northern Hungary	70	3,000	0	0	0	2.04	0
Southern Great Plain	200	2,000	1,000	5,500	0.3	4.41	83

parameters "o" and "u" are 600 and 3,000 (million €), respectively, with corresponding costs of around 13.85 million €. Furthermore, for insurance the attachment level will be 1,000 and the exit point 5,500 with a proportional factor of 0.7. Already indicated through the loading factors, insurance is much more expensive with annual costs to be 170 million €. Table 16.3 shows the parameters calculated for each important region (i.e., where losses increase) and the corresponding costs for mitigation and insurance.

According to our calculations Northern Hungary will be advised to invest in mitigation to tackle the increases in the more frequent losses, while the Southern Great Plain should consider employing both mitigation and insurance instruments. These are less costly due to the absolute smaller increase in risk compared to the Northern Great Plain.

Summarizing, this illustrative example demonstrates the value of a risk layered approach in considering adaptation measures as they are differentiated between risk reduction and insurance. While the results are only indicative, the methodology shows how quantitative modeling approaches can be useful for analyzing pre-disaster risk management options.

16.5 Conclusion

This chapter presented current and future direct flood loss estimates for Hungary and the Tisza region based on a risk modelling approach (CATSIM, Chap. 8). It was found that the Northern Great Plain, in particular, as well as the Southern Great Plain, have larger estimated losses for 2071–2100 (assuming equivalent vulnerability and exposure as today). We demonstrated how a risk layering approach that distinguishes between more frequent and less frequent impacts provides insights on two key risk management strategies: risk reduction and risk transfer.

Several limitations of the study, most of them data related, reduce its applicability. The analysis did not include land-use and demographic dynamics, nor did it account for changes in vulnerability. It also did not take into account of lives lost from flood events. Furthermore, it would be important to examine different scenarios when investigating the impacts of climate change. While the necessary data will likely not be forthcoming in the near future, there is value in examining risk-management issues from this broad-brush perspective. The results can provide insights on adaptation strategies that are robust under changing future conditions and given budget constraints.

References

Alcamo J, Moreno JM, Nováky B, Bindi M, Corobov R, Devoy RJN et al (2007) Europe. Climate change 2007: impacts, adaptation and vulnerability. Contribution of working group II to the fourth assessment report of the intergovernmental panel on climate change. In: Parry ML, Canziani OF, Palutikof JP, van der Linden PJ, Hanson CE (eds) Climate change 2007: impacts, adaptation and vulnerability. Cambridge University Press, Cambridge, pp 541–580

Barredo JI, Lavalle C, De Roo A (2005) European flood risk mapping. EC DGJRC, 2005 S.P. I.05.151.EN

Compton KL, Faber R, Ermolieva TY, Linnerooth-Bayer J, Nachtnebel H-P (2009) Uncertainty and disaster risk management: modeling the flash flood risk to Vienna and its subway system. IIASA research report, RR-09-002, Laxenburg, Austria

Dankers R, Feyen L (2008) Climate change impact on flood hazard in Europe: an assessment based on high resolution climate simulations. J Geophys Res 113:D19105. doi:10.1029/2007JD009719

De Roo A (1998) Modelling runoff and sediment transport in catchments using GIS. GIS applications in hydrology. Hydrol Process 12:905–922

EEA (2000) CORINE land cover technical guide, European Environment Agency, Technical report 40. http://www.eea.europa.eu/publications/tech40add/page001.html

Grossi P, Kunreuther H (eds) (2005) Catastrophe modeling: a new approach to managing risk. Springer, New York

Hirabayashi Y, Kanae S, Emori S, Oki T, Kimoto M (2008) Global projections of changing risks of floods and droughts in a changing climate. Hydrol Sci J 53:754–773

Hochrainer S (2006) Macroeconomic risk management against natural disasters. German University Press, Wiesbaden

Hochrainer S, Mechler R (2009) Report on Europe's financial and economic vulnerability to meteorological extremes. A.2.3. Final deliverable, ADAM, Brussels, Belgium

Hochrainer S, Lugeri N, Radziejewski M (2012) Up-scaling of impact dependent loss distributions: a hybrid-convolution approach (forthcoming)

IPCC (2000) IPCC special report: emissions scenarios. Summary for policymakers. UNEP/WPO, Intergovernmental Panel on Climate Change. http://www.ipcc.ch/pdf/special-reports/spm/sres-en.pdf

Kovacs GZ (2006) Flood control in Hungary – position paper. In: Van Alphen J, van Beek E, Taal M (eds) Floods, from defence to management. Taylor & Francis Group, London, pp 157–163

Kundzewicz ZW, Mechler R (2010) Assessing adaptation to extreme weather events in Europe. Mitig Adapt Strat Glob Change 15:612–620

Kundzewicz ZB, Lugeri B, Dankers R, Hirabayashi Y, Doell P, Pinskwar I, Dysarz T, Hochrainer S, Matczak P (2010) Assessing river flood risk and adaptation in Europe – review of projections for the future. Mitig Adapt Strat Glob Change 15(7):641–656

Lindell MK, Perry RW (2004) Communicating environmental risk in multiethnic communities. Sage, Thousand Oaks

Linnerooth-Bayer J, Quijano-Evans S, Löfstedt R, Elahi S (2001) The uninsured elements of natural catastrophic losses: seven case studies of earthquake and flood disasters, summary report. Project funded by the UK Tsunami fund, IIASA, Laxenburg, Austria

Lugeri N, Lavalle C, Hochrainer S, Bindi M, Moriondo M (2009) An assessment of weather-related risks in Europe. A.2.1. Final deliverable. ADAM, Brussels, Belgium

Lugeri N, Kundzewicz ZB, Genovese E, Hochrainer S, Radziejewski M (2010) River flood risk and adaptation in Europe – assessment of the present status. Mitig Adapt Strat Glob Change 15 (7):621–640

Mechler R (2004) Natural disaster risk management and financing disaster losses in developing countries. Verlag Versicherungswirtschaft GmbH, Karlsruhe

Mechler R, Hochrainer S, Pflug G, Lotsch A, Williges K (2009) Assessing the financial vulnerability to climate-related natural hazards. Policy research working paper. 5232, World Bank, Washington, DC

MMC (2005) Natural hazard mitigation saves: an independent study to assess the future savings from mitigation activities, vol 2, Study documentation. Multihazard Mitigation Council, Washington, DC

Parry ML, Canziani OF, Palutikof JP, van der Linden PJ, Hanson CE (eds) (2007) Climate change 2007: impacts adaptation and vulnerability. Contribution of working group II to the fourth assessment report of the intergovernmental panel of climate change. Cambridge University Press, Cambridge

Pflug G, Römisch W (2007) Modelling, measuring and managing risk. World Scientific, Singapore

Rescher N (1983) Risk: a philosophical introduction to the theory of risk evaluation and management. University Press of America, Washington, DC

Index

A
Actuarial (analysis, approach), 14, 16, 26, 32–33, 36, 45, 54, 55
Adaptive (capacity, management, measure, process), 81, 100, 129, 172, 182, 246
  Monte Carlo Optimization (*see* Monte Carlo simulation)
Agriculture, agricultural (activity, damage, exposure), 59, 74, 82, 150–151, 163–164, 166, 167, 173–174, 176, 191–193, 200
Aid (governmental, external, foreign), 63, 69, 86, 96, 108–109, 120, 125, 126, 133, 137, 149, 160, 166, 207, 211. *See also* Financial aid
Aleatory uncertainty, 10, 15–17, 26
Attachment point, layer (for risk transfer), 23, 127, 137, 274, 277

B
Bac Hung Hai polder, region, 11, 53–54, 59, 63
Basin (retention, detention), 10, 13, 17, 19, 21, 24, 26, 27, 174
  River (*see* Tisza river; Vienna river)
Bereg, 172, 176, 181, 184–196, 220
Biodiversity, 9, 122, 174, 183, 200
Bodrogköz, 176, 183–194, 196
Bond (catastrophe), 8, 23, 35, 100, 109, 112, 116, 126, 132, 133, 137, 141, 152, 154, 159, 160, 247
Borrowing (contingent, ex-post), 23, 24, 26, 31, 35, 42, 43, 56, 63, 64, 103, 104, 108–113, 115, 124, 132, 133, 147, 154, 159–160, 181, 185, 189, 194–197, 247, 248, 254

B
Business (interruption. losses), 31, 95, 115, 124, 161, 206

C
Capital (accumulation, movement, exposed, reserve etc.), 4, 5, 29, 35, 37, 48, 74, 79, 80, 82, 97, 99, 102, 104–106, 108, 109, 113–115, 120–121, 127, 130–131, 133, 137–139, 155–159, 162–163, 167, 173, 246–248, 253, 260
Caribbean, 14
Caribbean Catastrophe Risk Insurance Facility (CCRIF), 8
Cascading events, failures, 6, 30, 249
Case study, 1, 11, 14, 29, 31, 32, 34, 36, 43–50, 53, 58–60, 76, 82, 84–86, 90, 140, 181–197, 231, 255
Catastrophe fund. *See* Fund
Catastrophe insurance. *See* Insurance
Catastrophe model(s), modeling, 3–11, 13–14, 17, 20, 29, 31, 33–37, 50, 76, 81–84, 113, 115, 118, 122, 165, 171–172, 176–177, 199, 201, 215, 217–218, 245, 247–251, 261, 264–265
CATastrophe SIMulation (CATSIM), 8, 100, 119–141, 145–167, 264, 265, 278
Catchment models, 18, 267
CGE. *See* Computable general equilibrium (CGE)
China, 75
Climate change, 4, 5, 8, 18, 30, 120, 121, 131, 171, 172, 174, 177, 182, 236, 263–264, 269–272, 277–278
  adaptation, 263, 264, 272, 273, 277, 278

A. Amendola et al. (eds.), *Integrated Catastrophe Risk Modeling: Supporting Policy Processes*, Advances in Natural and Technological Hazards Research 32, DOI 10.1007/978-94-007-2226-2, © Springer Science+Business Media Dordrecht 2013

CPSIA information can be obtained
at www.ICGtesting.com
Printed in the USA
LVOW01s1500241016
510063LV00002B/17/P